CAMBRIDGE LIBRARY COLLECTION

Books of enduring scholarly value

Technology

The focus of this series is engineering, broadly construed. It covers technological innovation from a range of periods and cultures, but centres on the technological achievements of the industrial era in the West, particularly in the nineteenth century, as understood by their contemporaries. Infrastructure is one major focus, covering the building of railways and canals, bridges and tunnels, land drainage, the laying of submarine cables, and the construction of docks and lighthouses. Other key topics include developments in industrial and manufacturing fields such as mining technology, the production of iron and steel, the use of steam power, and chemical processes such as photography and textile dyes.

The Life of Thomas Telford, Civil Engineer

This biography of the civil engineer Thomas Telford (1757–1834) was published in 1867 by Samuel Smiles (1812–1904), the author of *Self-Help* and of other biographies of engineers and innovators. Smiles had already written about Telford's life and achievements in Volume 2 of his *Lives of the Engineers* (which is also reissued in this series), but in returning to the topic he adds to this new edition an introductory section (taken from Volume 1 of *Lives of the Engineers*) on the history of roads in Britain, from prehistoric trackways, via the Romans, to the modern road-building system pioneered by John Metcalf (the extraordinary 'Blind Jack of Knaresborough') and Telford himself. This illustrated work gives engaging accounts from earlier writers of the perils of road travel, and also deals in detail with Telford's own career as a builder of roads, bridges and canals.

Cambridge University Press has long been a pioneer in the reissuing of out-of-print titles from its own backlist, producing digital reprints of books that are still sought after by scholars and students but could not be reprinted economically using traditional technology. The Cambridge Library Collection extends this activity to a wider range of books which are still of importance to researchers and professionals, either for the source material they contain, or as landmarks in the history of their academic discipline.

Drawing from the world-renowned collections in the Cambridge University Library and other partner libraries, and guided by the advice of experts in each subject area, Cambridge University Press is using state-of-the-art scanning machines in its own Printing House to capture the content of each book selected for inclusion. The files are processed to give a consistently clear, crisp image, and the books finished to the high quality standard for which the Press is recognised around the world. The latest print-on-demand technology ensures that the books will remain available indefinitely, and that orders for single or multiple copies can quickly be supplied.

The Cambridge Library Collection brings back to life books of enduring scholarly value (including out-of-copyright works originally issued by other publishers) across a wide range of disciplines in the humanities and social sciences and in science and technology.

The Life of Thomas Telford, Civil Engineer

With an Introductory History of Roads and Travelling in Great Britain

Samuel Smiles

CAMBRIDGE
UNIVERSITY PRESS

University Printing House, Cambridge, CB2 8BS, United Kingdom

Cambridge University Press is part of the University of Cambridge.
It furthers the University's mission by disseminating knowledge in the pursuit of
education, learning and research at the highest international levels of excellence.

www.cambridge.org
Information on this title: www.cambridge.org/9781108067898

This edition first published 1867
This digitally printed version 2014

ISBN 978-1-108-06789-8 Paperback

THOMAS TELFORD, C.E., F.R.S.

Frontispiece

THE LIFE

OF

THOMAS TELFORD,

CIVIL ENGINEER.

WITH AN INTRODUCTORY HISTORY OF ROADS AND TRAVELLING IN GREAT BRITAIN.

BY SAMUEL SMILES,

AUTHOR OF 'SELF-HELP,' 'INDUSTRIAL BIOGRAPHY,' ETC.

"Let us travel, and wherever we find no facility for travelling from a city to a town, from a village to a hamlet, we may pronounce the people to be barbarous."—*Abbé Raynal.*

"The opening up of the internal communications of a country is undoubtedly the first and most important element of its growth in commerce and civilization."—*Richard Cobden.*

A NEW EDITION.

LONDON:
JOHN MURRAY, ALBEMARLE STREET.
1867.

WORKS BY THE SAME AUTHOR.

THE HUGUENOTS: their Settlements, Churches, and Industries, in England and Ireland. 8vo.

LIVES OF THE ENGINEERS: illustrated with 9 Portraits on Steel, and 342 Woodcuts. 4 vols. 8vo. 21*s*. each.

VOL. I. VERMUYDEN; MYDDELTON; BRINDLEY.
„ II. SMEATON; RENNIE; TELFORD.
„ III. GEORGE AND ROBERT STEPHENSON.
„ IV. BOULTON AND WATT.

POPULAR EDITIONS. *Post 8vo. 6s. each.*

SELF-HELP. 75*th Thousand.*
INDUSTRIAL BIOGRAPHY. 15*th Thousand.*
LIVES OF THE STEPHENSONS. 30*th Thousand.*
LIVES OF BRINDLEY AND THE EARLY ENGINEERS.

LONDON: PRINTED BY W. CLOWES AND SONS, STAMFORD STREET, AND CHARING CROSS.

PREFACE.

The present is a revised and in some respects enlarged edition of the 'Life of Telford,' originally published in the 'Lives of the Engineers,' to which is prefixed an account of the early roads and modes of travelling in Britain.

From this volume, read in connection with the Lives of George and Robert Stephenson, in which the origin and extension of Railways is described, an idea may be formed of the extraordinary progress which has been made in opening up the internal communications of this country during the last century.

Among the principal works executed by Telford in the course of his life, were the great highways constructed by him in North Wales and the Scotch Highlands, through districts formerly almost inaccessible, but which are now as easily traversed as any English county.

By means of these roads, and the facilities afforded by railways, the many are now enabled to visit with ease and comfort magnificent mountain scenery, which before was only the costly privilege of the few; at the same time that their construction has exercised a most beneficial influence on the population of the districts themselves.

The Highland roads, which were constructed with the active assistance of the Government, and were maintained partly at the public expense until within the last few years, had the effect of stimulating industry, improving agriculture, and converting a turbulent because unemployed population into one of the most loyal and well-conditioned in the empire;—the policy thus adopted with reference to the Highlands, and the beneficial results which have flowed from it, affording the strongest encouragement to Government in dealing in like manner with the internal communications of Ireland.

While the construction of the Highland roads was in progress, the late Robert Southey, poet laureate, visited the Highlands in company with his friend the engineer, and left on record an interesting account of his visit, in a manuscript now in the possession of Robert Rawlinson, C.E., to whom we are indebted for the extracts which are made from it in the present volume.

London, October, 1867.

CONTENTS.

EARLY ROADS AND MODES OF TRAVELLING.

CHAPTER I.

OLD ROADS.

CHAPTER II.

EARLY MODES OF CONVEYANCE.

CHAPTER III.

MANNERS AND CUSTOMS INFLUENCED BY THE STATE OF THE ROADS.

CHAPTER IV.

ROADS AND TRAVELLING IN SCOTLAND IN THE LAST CENTURY.

CHAPTER V.

ROADS AND TRAVELLING IN ENGLAND TOWARDS THE END OF LAST CENTURY.

CHAPTER VI.

JOHN METCALF, ROAD-MAKER.

LIFE OF THOMAS TELFORD.

CHAPTER I.

ESKDALE.

CHAPTER II.

LANGHOLM — TELFORD A STONEMASON.

CHAPTER III.

WORKING MASON IN LONDON, AND FOREMAN AT PORTSMOUTH.

CHAPTER IV.

SURVEYOR FOR THE COUNTY OF SALOP.

CHAPTER V.

FIRST EMPLOYMENT AS ENGINEER.

CHAPTER VI.

THE ELLESMERE CANAL.

CHAPTER VII.

Iron and other Bridges.

CHAPTER VIII.

Highland Roads and Bridges.

CHAPTER IX.

Telford's Scotch Harbours.

CHAPTER X.

Caledonian and other Canals.

CHAPTER XI.

Telford as a Road-maker.

CHAPTER XII.

The Menai and Conway Suspension Bridges.

CHAPTER XIII.

Docks, Drainage, and Bridges.

CHAPTER XIV.

Southey's Tour in the Highlands.

CHAPTER XV.

Telford's later Years — Death and Character.

LIST OF ILLUSTRATIONS.

ANCIENT CAUSEWAY, NEAR WHITBY.

By PERCIVAL SKELTON.

Page 1.

EARLY ROADS

AND

MODES OF TRAVELLING.

CHAPTER I.

OLD ROADS.

ROADS have in all times been among the most influential agencies of society; and the makers of them, by enabling men readily to communicate with each other, have properly been regarded as among the most effective pioneers of civilization.

Roads are literally the pathways not only of industry, but of social and national intercourse. Wherever a line of communication between men is formed, it renders commerce practicable; and, wherever commerce penetrates, it creates a civilization and leaves a history.

Roads place the city and the town in connection with the village and the farm, open up markets for field produce, and provide outlets for manufactures. They enable the natural resources of a country to be developed, facilitate travelling and intercourse, break down local jealousies, and in all ways tend to bind together society and bring out fully that healthy spirit of industry which is the life and soul of every nation.

The road is so necessary an instrument of social well-being, that in every new colony it is one of the first things thought of. First roads, then commerce, institutions, schools, churches, and newspapers. The new country, as well as the old, can only be effectually "opened up," as the common phrase is, by roads; and until these are made, it is virtually closed.

Freedom itself cannot exist without free communication,—every limitation of movement on the part of the members of society amounting to a positive abridgment of their personal liberty. Hence roads, canals, and railways, by providing the greatest possible facilities for locomotion and information, are essential for the freedom of all classes, of the poorest as well as the richest.

By bringing the ends of a kingdom together, they reduce the inequalities of fortune and station, and, by equalizing the price of commodities, to that extent they render them accessible to all. Without their assistance, the concentrated populations of our large towns could neither be clothed nor fed; but by their instrumentality an immense range of country is brought as it were to their very doors, and the sustenance and employment of large masses of people become comparatively easy.

In the raw materials required for food, for manufactures, and for domestic purposes, the cost of transport necessarily forms a considerable item; and it is clear that the more this cost can be reduced by facilities of communication, the cheaper these articles become, and the more they are multiplied and enter into the consumption of the community at large.

Let any one imagine what would be the effect of closing the roads, railways, and canals of England. The country would be brought to a dead lock, employment would be restricted in all directions, and a large proportion of the inhabitants concentrated in the large towns must at certain seasons inevitably perish of cold and hunger.

In the earlier periods of English history, roads were of comparatively less consequence. While the population was thin and scattered, and men lived by hunting and pastoral pursuits, the track across the down, the heath, and the moor, sufficiently answered their purpose. Yet even in those districts unencumbered with wood, where the first settlements were made—as on the downs of Wiltshire, the moors of Devonshire, and the wolds of Yorkshire —stone tracks were laid down by the tribes between one village and another. We have given, at the beginning of this chapter, a representation of one of those ancient trackways still existing in the neighbourhood of Whitby, in Yorkshire; and there are many of the same description to be met with in other parts of England. In some districts they are called trackways or ridgeways, being narrow causeways usually following the natural ridge of the country, and probably serving in early times as local boundaries. On Dartmoor they are constructed of stone blocks, irregularly laid down on the surface of the ground, forming a rude causeway of about five or six feet wide.

The Romans, with many other arts, first brought into England the art of road-making. They thoroughly understood the value of good roads, regarding them as the essential means for the maintenance of their empire in the first instance, and of social prosperity in the next. It was their roads, as well as their legions, that made them masters of the world; and the pickaxe, not less than the sword, was the ensign of their dominion. Wherever they went, they opened up the communications of the countries they subdued, and the roads which they made were among the best of their kind. They were skilfully laid out and solidly constructed. For centuries after the Romans left England, their roads continued to be the main highways of internal communication, and their remains are to this day to be traced in many parts of the country. Settlements were made and towns sprang up along the old "streets;" and the numerous Stretfords, Stratfords, and towns ending in "le-street"—as Ardwick-le-street, in Yorkshire, and Chester-le-street, in Durham—mostly mark the direction of these ancient lines of road. There are also numerous Stanfords, which were so called because they bordered the raised military roadways of the Romans, which ran direct between their stations.

The last-mentioned peculiarity of the roads constructed by the Romans, must have struck many observers. Level does not seem to have been of consequence, compared with directness. This peculiarity is supposed to have originated in an imperfect knowledge of mechanics; for the Romans do not appear to have been acquainted with the moveable joint in wheeled carriages. The carriage-body rested solid upon the axles, which in four-wheeled vehicles were rigidly parallel with each other. Being unable readily to turn a bend in the road, it has been concluded that for this reason all the great Roman highways were constructed in as straight lines as possible.

On the departure of the Romans from Britain, most of the roads constructed by them were allowed to fall into decay, on which the forest and the waste gradually resumed their dominion over them, and the highways of England became about the worst in Europe. We find, however, that numerous attempts were made in early times to preserve the ancient ways and enable a communication to be maintained between the metropolis and the rest of the country, as well as between one market town and another.

The state of the highways may be inferred from the character of the legislation applying to them. One of the first laws on the subject was passed in 1285, directing that all bushes and trees along the roads leading from one market to another should be cut down for two hundred feet on either side, to prevent robbers lurking therein;*

* Brunetto Latini, the tutor of Dante, describes a journey made by him from London to Oxford about the end of the thirteenth century, resting by the way at Shirburn Castle. He says, "Our journey from London to Oxford was, with some difficulty and danger, made in two days; for the roads are bad, and we had to climb hills of hazardous ascent, and which to descend are equally perilous. We passed through many woods, considered here as dangerous places, as they are infested with robbers, which indeed is the case with most of the roads in England. This is a circumstance connived at by the neighbouring barons, on consideration of sharing in the booty, and of

but nothing was proposed for amending the condition of the ways themselves. In 1346, Edward III. authorised the first toll to be levied for the repair of the roads leading from St. Giles's-in-the-Fields to the village of Charing (now Charing Cross), and from the same quarter to near Temple Bar (down Drury Lane), as well as the highway then called Perpoole (now Gray's Inn Lane). The footway at the entrance of Temple Bar was interrupted by thickets and bushes, and in wet weather was almost impassable. The roads further west were so bad that when the sovereign went to Parliament faggots were thrown into the ruts in King-street, Westminster, to enable the royal cavalcade to pass along.

In Henry VIII.'s reign, several remarkable statutes were passed relating to certain worn-out and impracticable roads in Sussex and the Weald of Kent. From the earliest of these, it would appear that when the old roads were found too deep and miry to be passed, they were merely abandoned and new tracks struck out. After describing "many of the wayes in the wealds as so depe and noyous by wearyng and course of water and other occasions that people cannot have their carriages or passages by horses uppon or by the same but to their great paynes, perill and jeopardie," the Act provided that owners of land might, with the consent of two justices and twelve discreet men of the hundred, lay out new roads and close up the old ones. Another Act passed in the same reign, related to the repairs of bridges and of the highways at the ends of bridges.

But as these measures were for the most part merely permissive, they could have had but little practical effect in improving the communications of the kingdom. In the

these robbers serving as their protectors on all occasions, personally, and with the whole strength of their band. However, as our company was numerous, we had less to fear. Accordingly, we arrived the first night at Shirburn Castle, in the neighbourhood of Watlington, under the chain of hills over which we passed at Stokenchurch." This passage is given in Mr. Edward's work on 'Libraries' (p. 328), as supplied to him by Lady Macclesfield.

reign of Philip and Mary (in 1555), an Act was passed providing that each parish should elect two surveyors of highways to see to the maintenance of their repairs by compulsory labour, the preamble reciting that "highwaies are now both verie noisome and tedious to travell in, and dangerous to all passengers and cariages;" and to this day parish and cross roads are maintained on the principle of Mary's Act, though the compulsory labour has since been commuted into a compulsory tax.

In the reigns of Elizabeth and James, other road Acts were passed; but, from the statements of contemporary writers, it would appear that they were followed by very little substantial progress, and travelling continued to be attended with many difficulties. Even in the neighbourhood of the metropolis, the highways were in certain seasons scarcely passable. The great Western road into London was especially bad, and about Knightsbridge, in winter, the traveller had to wade through deep mud. Wyatt's men entered the city by this approach in the rebellion of 1554, and were called the "draggle-tails" because of their wretched plight. The ways were equally bad as far as Windsor, which, in the reign of Elizabeth, is described by Pote, in his history of that town, as being "not much past half a day's journeye removed from the flourishing citie of London."

At a greater distance from the metropolis, the roads were still worse. They were in many cases but rude tracks across heaths and commons, as furrowed with deep ruts as ploughed fields; and in winter to pass along one of them was like travelling in a ditch. The attempts made by the adjoining occupiers to mend them, were for the most part confined to throwing large stones into the bigger holes to fill them up. It was easier to allow new tracks to be made than to mend the old ones. The land of the country was still mostly unenclosed, and it was possible, in fine weather, to get from place to place, in one way or another, with the help of a guide. In the absence of bridges, guides were necessary to point out the safest fords as well as to pick out the least miry tracks. The most frequented lines of road were struck out from time to time by the drivers of pack-horses, who, to avoid the bogs and sloughs, were usually careful to keep along the higher grounds; but, to prevent those horsemen who departed from the beaten track being swallowed up in quagmires, beacons were erected to warn them against the more dangerous places.*

In some of the older-settled districts of England, the old roads are still to be traced in the hollow Ways or Lanes, which are to be met with, in some places, eight and ten feet deep. They were horse-tracks in summer, and rivulets in winter. By dint of weather and travel, the earth was gradually worn into these deep furrows, many of which, in Wilts, Somerset, and Devon, represent the tracks of roads as old as, if not older than, the Conquest. When the ridge-ways of the earliest settlers on Dartmoor, above alluded to, were abandoned, the tracks were formed through the valleys, but the new roads were no better than the old ones. They were narrow and deep, fitted only for a horse passing along laden with its crooks, as so graphically described in the ballad of "The Devonshire Lane." †

* See Ogilvy's 'Britannia Depicta,' the traveller's ordinary guide-book between 1675 and 1717, as Bradshaw's Railway Time-book is now. The Grand Duke Cosmo, in his 'Travels in England in 1669,' speaks of the country between Northampton and Oxford as for the most part unenclosed and uncultivated, abounding in weeds. From Ogilby's fourth edition, published in 1749, it appears that the roads in the midland and northern districts of England were still, for the most part, entirely unenclosed.

† This ballad is so descriptive of the old roads of the south-west of England that we are tempted to quote it at length. It was written by the Rev. John Marriott, sometime vicar of Broadclist, Devon; and Mr. Rowe, vicar of Crediton, says, in his 'Perambulation of Dartmoor,' that he can readily imagine the identical lane near Broadclist, leading towards Poltemore, which might have *sat* for the portrait.

In a Devonshire lane, as I trotted along
T'other day, much in want of a subject for song,
Thinks I to myself, half-inspired by the rain,
Sure marriage is much like a Devonshire lane.

Similar roads existed until recently in the immediate neighbourhood of Birmingham, now the centre of an immense traffic. The sandy soil was sawn through, as it were, by generation after generation of human feet, and by pack-horses, helped by the rains, until in some places the tracks were as much as from twelve to fourteen yards deep; one of these, partly filled up, retaining to this day the name of Holloway Head. In the neighbourhood of London there was also a Hollow way, which now gives its name to a populous metropolitan parish. Hagbush Lane was another of such roads. Before the formation of the Great North Road, it was one of the principal bridle-paths leading from London to the northern parts of England; but it was so narrow as barely to afford passage for more than a single horseman, and so deep that the rider's head was beneath the level of the ground on either side.

In the first place 'tis long, and when once you are in it,
It holds you as fast as a cage does a linnet;
For howe'er rough and dirty the road may be found,
Drive forward you must, there is no turning round.

But tho' 'tis so long, it is not very wide,
For two are the most that together can ride;
And e'en then, 'tis a chance but they get in a pother,
And jostle and cross and run foul of each other.

Oft poverty meets them with mendicant looks,
And care pushes by them with dirt-laden crooks;
And strife's grazing wheels try between them to pass,
And stubbornness blocks up the way on her ass.

Then the banks are so high, to the left hand and right,
That they shut up the beauties around them from sight;
And hence, you'll allow, 'tis an inference plain,
That marriage is just like a Devonshire lane.

But thinks I, too, these banks, within which we are pent,
With bud, blossom, and berry, are richly besprent;
And the conjugal fence, which forbids us to roam,
Looks lovely, when deck'd with the comforts of home.

In the rock's gloomy crevice the bright holly grows;
The ivy waves fresh o'er the withering rose,
And the ever-green love of a virtuous wife
Soothes the roughness of care, cheers the winter of life.

Then long be the journey, and narrow the way,
I'll rejoice that I've seldom a turnpike to pay;
And whate'er others say, be the last to complain,
Though marriage is just like a Devonshire lane.

The roads of Sussex long preserved an infamous notoriety. Chancellor Cowper, when a barrister on circuit, wrote to his wife in 1690, that "the Sussex ways are bad and ruinous beyond imagination. I vow 'tis melancholy consideration that mankind will inhabit such a heap of dirt for a poor livelihood. The country is a sink of about fourteen miles broad, which receives all the water that falls from two long ranges of hills on both sides of it, and not being furnished with convenient draining, is kept moist and soft by the water till the middle of a dry summer, which is only able to make it tolerable to ride for a short time."

It was almost as difficult for old persons to get to church in Sussex during winter as it was in the Lincoln Fens, where they were rowed thither in boats. Fuller saw an old lady being drawn to church in her own coach by the aid of six oxen. The Sussex roads were indeed so bad as to pass into a by-word. A contemporary writer says, that in travelling a slough of extraordinary miryness, it used to be called "the Sussex bit of the road;" and he satirically alleged that the reason why the Sussex girls were so long-limbed was because of the tenacity of the mud in that county; the practice of pulling the foot out of it "by the strength of the ancle" tending to stretch the muscle and lengthen the bone! *

But the roads in the immediate neighbourhood of London long continued almost as bad as those in Sussex. Thus, when the poet Cowley retired to Chertsey, in 1665, he wrote to his friend Sprat to visit him, and, by way of encouragement, told him that he might sleep the first night at Hampton town; thus occupying two days in the performance of a journey of twenty-two miles in the immediate neighbourhood of the metropolis. As late as 1736 we find Lord Hervey, writing from Kensington, complaining that "the road between this place and London is grown so

---

* '*Iter Sussexiense.*' By Dr. John Burton.

infamously bad that we live here in the same solitude as we would do if cast on a rock in the middle of the ocean; and all the Londoners tell us that there is between them and us an impassable gulf of mud."

Nor was the mud any respecter of persons; for we are informed that the carriage of Queen Caroline could not, in bad weather, be dragged from St. James's Palace to Kensington in less than two hours, and occasionally the royal coach stuck fast in a rut, or was even capsized in the mud. About the same time, the streets of London themselves were little better, the kennel being still permitted to flow in the middle of the road, which was paved with round stones,—flag-stones for the convenience of pedestrians being as yet unknown. In short, the streets in the towns and the roads in the country were alike rude and wretched,—indicating a degree of social stagnation and discomfort which it is now difficult to estimate, and almost impossible to describe.

CHAPTER II.

EARLY MODES OF CONVEYANCE.

SUCH being the ancient state of the roads, the only practicable modes of travelling were on foot and on horseback. The poor walked and the rich rode. Kings rode and Queens rode. Judges rode circuit in jack-boots. Gentlemen rode and robbers rode. The Bar sometimes walked and sometimes rode. Chaucer's ride to Canterbury will be remembered as long as the English language lasts. Hooker rode to London on a hard-paced nag, that he might be in time to preach his first sermon at St. Paul's. Ladies rode on pillions, holding on by the gentleman or the serving-man mounted before.

Shakespeare incidentally describes the ancient style of travelling among the humbler classes in his 'Henry IV.' * The party, afterwards set upon by Falstaff and his companions, bound from Rochester to London, were up by two in the morning, expecting to perform the journey of thirty miles by close of day, and to get to town "in time to go to bed with a candle." Two are carriers, one of whom has "a gammon of bacon and two razes of ginger, to be delivered as far as Charing Cross;" the other has his panniers full of turkeys. There is also a franklin of Kent, and another, "a kind of auditor," probably a tax-collector, with several more, forming in all a company of eight or ten, who travel together for mutual protection. Their robbery on Gad's Hill, as painted by Shakespeare, is but a picture, by no means exaggerated, of the adventures and dangers of the road at the time of which he wrote.

* King Henry the Fourth (Part I.), Act II. Scene 1.

Distinguished personages sometimes rode in horse-litters; but riding on horseback was generally preferred. Queen Elizabeth made most of her journeys in this way,* and when she went into the City she rode on a pillion behind her Lord Chancellor. The Queen, however, was at length provided with a coach, which must have been a very remarkable machine. This royal vehicle is said to have been one of the first coaches used in England, and it was introduced by the Queen's own coachman, one Boomen, a Dutchman. It was little better than a cart without springs, the body resting solid upon the axles. Taking the bad roads and ill-paved streets into account, it must have been an excessively painful means of conveyance. At one of the first audiences which the Queen gave to the French ambassador in 1568, she feelingly described to him "the aching pains she was suffering in consequence of having been knocked about in a coach which had been driven a little too fast, only a few days before."†

Such coaches were at first only used on state occasions. The roads, even in the immediate neighbourhood of London, were so bad and so narrow that the vehicles could not be taken into the country. But, as the roads became improved, the fashion of using them spread. When the aristocracy removed from the City to the western parts of the metropolis, they could be better accommodated, and in course of time they became gradually adopted. They were still, however, neither more nor less than waggons, and, indeed, were called by that name; but wherever they went they excited great wonder. It is related of "that valyant knyght Sir Harry Sidney," that on a certain day in

* Part of the riding road along which the Queen was accustomed to pass on horseback between her palaces at Greenwich and Eltham is still in existence, a little to the south of Morden College, Blackheath. It winds irregularly through the fields, broad in some places, and narrow in others. Probably it is very little different from what it was when used as a royal road. It is now very appropriately termed "Muddy Lane."

† 'Dépêches de La Mothe Fénelon,' 8vo., 1838. Vol. i. p. 27.

the year 1583 he entered Shrewsbury in his waggon, "with his Trompeter blowynge, verey joyfull to behold and see." *

From this time the use of coaches gradually spread, more particularly amongst the nobility, superseding the horse-litters which had till then been used for the conveyance of ladies and others unable to bear the fatigue of riding on horseback. The first carriages were heavy and lumbering: and upon the execrable roads of the time they went pitching over the stones and into the ruts, with the pole dipping and rising like a ship in a rolling sea. That they had no springs, is clear enough from the statement of Taylor, the water-poet—who deplored the introduction of carriages as a national calamity—that in the paved streets of London men and women were "tossed, tumbled, rumbled, and jumbled about in them." Although the road from London to Dover, along the old Roman Watling-street, was then one of the best in England, the French household of Queen Henrietta, when they were sent forth from the palace of Charles I., occupied four tedious days before they reached Dover.

But it was only a few of the main roads leading from the metropolis that were practicable for coaches; and on the occasion of a royal progress, or the visit of a lord-lieutenant, there was a general turn out of labourers and masons to mend the ways and render the bridges at least temporarily secure. Of one of Queen Elizabeth's journeys it is said:—"It was marvellous for ease and expedition, for such is the perfect evenness of the new highway that Her Majesty left the coach only once, while the hinds and the folk of a base sort *lifted it on with their poles.*"

Sussex long continued impracticable for coach travelling at certain seasons. As late as 1708, Prince George of Denmark had the greatest difficulty in making his way to Petworth to meet Charles VI. of Spain. "The last nine miles of the way," says the reporter, "cost us six hours to conquer them." One of the couriers in attendance complained that

* Nichols's 'Progresses,' vol. ii., 309.

during fourteen hours he never once alighted, except when the coach overturned, or stuck in the mud.

When the judges, usually old men and bad riders, took to going the circuit in their coaches, juries were often kept waiting until their lordships could be dug out of a bog or hauled out of a slough by the aid of plough-horses. In the seventeenth century, scarcely a Quarter Session passed without presentments from the grand jury against certain districts on account of the bad state of the roads, and many were the fines which the judges imposed upon them as a set-off against their bruises and other damages while on circuit.

THE OLD STAGE WAGGON. [By Louis Huard, after Rowlandson.]

For a long time the roads continued barely practicable for wheeled vehicles of the rudest sort, though Fynes Morison (writing in the time of James I.) gives an account of "carryers, who have long covered waggons, in which they carry passengers from place to place; but this kind of journeying," he says, "is so tedious, by reason they must take waggon very early and come very late to their innes, that none but women and people of inferior condition travel in this sort."

The waggons of which Morison wrote, made only from ten to fifteen miles in a long summer's day; that is, supposing them not to have broken down by pitching over the boulders laid along the road, or stuck fast in a quagmire, when they had to wait for the arrival of the next team of horses to help to drag them out. The waggon, however, continued to be adopted as a popular mode of travelling until late in the eighteenth century; and Hogarth's picture illustrating the practice will be remembered, of the cassocked parson on his lean horse, attending his daughter newly alighted from the York waggon.

A curious description of the state of the Great North Road, in the time of Charles II., is to be found in a tract published in 1675 by Thomas Mace, one of the clerks of Trinity College, Cambridge. The writer there addressed himself to the King, partly in prose and partly in verse, complaining greatly of the "wayes, which are so grossly foul and bad;" and suggesting various remedies. He pointed out that much ground "is now spoiled and trampled down in all wide roads, where coaches and carts take liberty to pick and chuse for their best advantages; besides, such sprawling and straggling of coaches and carts utterly confound the road in all wide places, so that it is not only unpleasurable, but extreme perplexin and cumbersome both to themselves and all horse travellers." It would thus appear that the country on either side of the road was as yet entirely unenclosed.

But Mace's principal complaint was of the "innumerable controversies, quarrellings, and disturbances" caused by the packhorse-men, in their struggles as to which convoy should pass along the cleaner parts of the road. From what he states, it would seem that these "disturbances, daily committed by uncivil, refractory, and rude Russian-like rake-shames, in contesting for the way, too often proved mortal, and certainly were of very bad consequences to many." He recommended a quick and prompt punishment in all such cases. "No man," said he, "should be pestered by giving the way (sometimes) to hundreds of pack-horses, panniers, whifflers (*i.e.* paltry fellows), coaches, waggons, wains, carts, or whatsoever others, which continually are very grievous to weary and loaden travellers; but more especially near the city and upon a market day, when, a man having travelled a long and tedious journey, his horse well nigh spent, shall sometimes be compelled to cross out of his way twenty times in one mile's riding, by the irregularity and peevish crossness of such-like whifflers and market women; yea, although their panniers be clearly empty, they will stoutly contend for the way with weary travellers, be they never so many, or almost of what quality soever." "Nay," said he further, "I have often known many travellers, and myself very often, to have been necessitated to stand stock still behind a standing cart or waggon, on most beastly and unsufferable deep wet wayes, to the great endangering of our horses, and neglect of important business: nor durst we adventure to stirr (for most imminent danger of those deep rutts, and unreasonable ridges) till it has pleased Mister Carter to jog on, which we have taken very kindly."

Mr. Mace's plan of road reform was not extravagant. He mainly urged that only two good tracks should be maintained, and the road be not allowed to spread out into as many as half-a-dozen very bad ones, presenting high ridges and deep ruts, full of big stones, and many quagmires. Breaking out into verse, he said—

"First let the wayes be regularly brought
To artificial form, and truly wrought;
So that we can suppose them firmly mended,
And in all parts the work well ended,
That not a stone 's amiss; but all compleat,
All lying smooth, round, firm, and wondrous neat."

After a good deal more in the same strain, he concluded—

> "There's only *one thing* yet worth thinking on—
> Which is, to put this work in execution." *

But we shall find that more than a hundred years passed before the roads throughout England were placed in a more satisfactory state than they were in the time of Mr. Mace.

The introduction of stage-coaches about the middle of the seventeenth century formed a new era in the history of travelling by road. At first they were only a better sort of waggon, and confined to the more practicable highways near London. Their pace did not exceed four miles an hour, and the jolting of the unfortunate passengers conveyed in them must have been very hard to bear. It used to be said of their drivers that they were "seldom sober, never civil, and always late."

The first mention of coaches for public accommodation is made by Sir William Dugdale in his Diary, from which it appears that a Coventry coach was on the road in 1659. But probably the first coaches, or rather waggons, were run between London and Dover, as one of the most practicable routes for the purpose. M. Sobrière, a French man of letters, who landed at Dover on his way to London in the time of Charles II., alludes to the existence of a stage-coach, but it seems to have had no charms for him, as the following passage will show: "That I might not," he says, "take post or be obliged to use the stage-coach, I went from Dover to London in a waggon. I was drawn by six

* The title of Mace's tract (British Museum) is "The Profit, Conveniency, and Pleasure for the whole nation: being a short rational Discourse lately presented to his Majesty concerning the Highways of England: their badness, the causes thereof, the reasons of these causes, the impossibility of ever having them well mended according to the old way of mending: but may most certainly be done, and for ever so maintained (according to this NEW WAY) substantially and with very much ease, &c., &c. Printed for the public good in the year 1675."

horses, one before another, and driven by a waggoner, who walked by the side of it. He was clothed in black, and appointed in all things like another St. George. He had a brave montrero on his head and was a merry fellow, fancied he made a figure, and seemed mightily pleased with himself."

Shortly after, coaches seem to have been running as far north as Preston in Lancashire, as appears by a letter from one Edward Parker to his father, dated November, 1663, in which he says, "I got to London on Saturday last; but my journey was noe ways pleasant, being forced to ride in the boote all the waye. Ye company yt came up with mee were persons of greate quality, as knights and ladyes. My journey's expense was 30*s*. This traval hath soe indisposed mee, yt I am resolved never to ride up againe in ye coatch."* These vehicles must, however, have considerably increased, as we find a popular agitation was got up against them. The Londoners nicknamed them "hell-carts;" pamphlets were written recommending their abolition; and attempts were even made to have them suppressed by Act of Parliament.

Thoresby occasionally alludes to stage-coaches in his Diary, speaking of one that ran between Hull and York in 1679, from which latter place he had to proceed by Leeds in the usual way on horseback. This Hull vehicle did not run in winter, because of the state of the roads; stage-coaches being usually laid up in that season like ships during Arctic frosts.† Afterwards, when a coach was put on between York and Leeds, it performed the journey of

* See 'Archæologia,' xx., pp. 443-76.

† "4th May, 1714. Morning: we dined at Grantham, had the annual solemnity (this being the first time the coach passed the road in May), and the coachman and horses being decked with ribbons and flowers, the town music and young people in couples before us; we lodged at Stamford, a scurvy, dear town. 5th May: had other passengers, which, though females, were more chargeable with wine and brandy than the former part of the journey, wherein we had neither; but the next day we gave them leave to treat themselves."—Thoresby's 'Diary,' vol. ii., 207.

twenty-four miles in eight hours;* but the road was so bad and dangerous that the travellers were accustomed to get out and walk the greater part of the way.

Thoresby often waxes eloquent upon the subject of his manifold deliverances from the dangers of travelling by coach. He was especially thankful when he had passed the ferry over the Trent in journeying between Leeds and London, having on several occasions narrowly escaped drowning there. Once, on his journey to London, some showers fell, which "raised the washes upon the road near Ware to that height that passengers from London that were upon that road swam, and a poor higgler was drowned, which prevented me travelling for many hours; yet towards evening we adventured with some country people, who conducted us over the meadows, whereby we missed the deepest of the Wash at Cheshunt, though we rode to the saddle-skirts for a considerable way, but got safe to Waltham Cross, where we lodged." †

On another occasion Thoresby was detained four days at Stamford by the state of the roads, and was only extricated from his position by a company of fourteen members of the House of Commons travelling towards London, who took him into their convoy, and set out on their way southward attended by competent guides. When the "waters were out," as the saying went, the country became closed, the roads being simply impassable. During the Civil Wars eight hundred horse were taken prisoners while sticking in the mud.‡ When rain fell, pedestrians, horsemen, and coaches alike came to a standstill until the roads dried again and enabled the wayfarers to proceed. Thus we read of two travellers stopped by the rains within a few miles of Oxford, who found it impossible to accomplish

* "May 22, 1708. At York. Rose between three and four, the coach being hasted by Captain Crome (whose company we had) upon the Queen's business, that we got to Leeds by noon; blessed be God for mercies to me and my poor family."—Thoresby s 'Diary,' vol. ii., 7.

† Thoresby's 'Diary,' vol. i., 295.

‡ Waylen's 'Marlborough.'

their journey in consequence of the waters that covered the country thereabout.

A curious account has been preserved of the journey of an Irish Viceroy across North Wales towards Dublin in 1685. The roads were so horrible that instead of the Viceroy being borne along in his coach, the coach itself had to be borne after him the greater part of the way. He was five hours in travelling between St. Asaph and Conway, a distance of only fourteen miles. Between Conway and Beaumaris he was forced to walk, while his wife was borne along in a litter. The carriages were usually taken to pieces at Conway and carried on the shoulders of stout Welsh peasants to be embarked at the Straits of Menai.

The introduction of stage-coaches, like every other public improvement, was at first regarded with prejudice, and had considerable obloquy to encounter. In a curious book published in 1673, entitled 'The Grand Concern of England Explained in several Proposals to Parliament,'* stage-coaches and caravans were denounced as among the greatest evils that had happened to the kingdom, being alike mischievous to the public, destructive to trade, and prejudicial to the landed interest. It was alleged that travelling by coach was calculated to destroy the breed of horses, and make men careless of good horsemanship,—that it hindered the training of watermen and seamen, and interfered with the public resources. The reasons given are curious. It was said that those who were accustomed to travel in coaches became weary and listless when they rode a few miles, and were unwilling to get on horseback—"not being able to endure frost, snow, or rain, or to *lodge in the fields;*" that to save their clothes and keep themselves clean and dry, people rode in coaches, and thus contracted an idle habit of body; that this was ruinous to trade, for that

* Reprinted in the 'Harleian Miscellany,' vol. viii., p. 547. Supposed to have been written by one John Gressot, of the Charterhouse.

"most gentlemen, before they travelled in coaches, used to ride with swords, belts, pistols, holsters, portmanteaus, and hat-cases, which, in these coaches, they have little or no occasion for: for, when they rode on horseback, they rode in one suit and carried another to wear when they came to their journey's end, or lay by the way; but in coaches a silk suit and an Indian gown, with a sash, silk stockings, and beaver-hats, men ride in, and carry no other with them, because they escape the wet and dirt, which on horseback they cannot avoid; whereas, in two or three journeys on horseback, these clothes and hats were wont to be spoiled; which done, they were forced to have new very often, and that increased the consumption of the manufactures and the employment of the manufacturers; which travelling in coaches doth in no way do."*

The writer of the same protest against coaches gives some idea of the extent of travelling by them in those days; for to show the gigantic nature of the evil he was contending against, he averred that between London and the three principal towns of York, Chester, and Exeter, not fewer

* There were other publications of the time as absurd (viewed by the light of the present day) as Gressot's. Thus, "A Country Tradesman," addressing the public in 1678, in a pamphlet entitled 'The Ancient Trades decayed, repaired again,—wherein are declared the several abuses that have utterly impaired all the ancient trades in the Kingdom,' urges that the chief cause of the evil had been the setting up of Stage-coaches some twenty years before. Besides the reasons for suppressing them set forth in the treatise referred to in the text, he says, "Were it not for them (the Stage-coaches), there would be more Wine, Beer, and Ale, drunk in the Inns than is now, which would be a means to augment the King's Custom and Excise. Furthermore they hinder the breed of horses in this kingdom [the same argument was used against Railways], because many would be necessitated to keep a good horse that keeps none now. Seeing, then, that there are few that are gainers by them, and that they are against the common and general good of the Nation, and are only a conveniency to some that have occasion to go to London, who might still have the same wages as before these coaches were in use, therefore *there is good reason they should be suppressed.* Not but that it may be lawful to hire a coach upon occasion, but that it should be unlawful only to keep a coach that should go *long journeys* constantly, from one stage or place to another, upon certain days of the week as they do now."—p. 27.

than eighteen persons, making the journey in five days, travelled by them weekly (the coaches running thrice in the week), and a like number back; "which come, in the whole, to eighteen hundred and seventy-two in the year." Another great nuisance, the writer alleged, which flowed from the establishment of the stage-coaches, was, that not only did the gentlemen from the country come to London in them oftener than they need, but their ladies either came with them or quickly followed them. "And when they are there they must be in the mode, have all the new fashions, buy all their clothes there, and go to plays, balls, and treats, where they get such a habit of jollity and a love to gaiety and pleasure, that nothing afterwards in the country will serve them, if ever they should fix their minds to live there again; but they must have all from London, whatever it costs."

Then there were the grievous discomforts of stage-coach travelling, to be set against the more noble method of travelling by horseback, as of yore. "What advantage is it to men's health," says the writer, waxing wroth, "to be called out of their beds into these coaches, an hour before day in the morning; to be hurried in them from place to place, till one hour, two, or three within night; insomuch that, after sitting all day in the summer-time stifled with heat and choked with dust, or in the winter-time starving and freezing with cold or choked with filthy fogs, they are often brought into their inns by torchlight, when it is too late to sit up to get a supper; and next morning they are forced into the coach so early that they can get no breakfast? What addition is this to men's health or business to ride all day with strangers, oftentimes sick, antient, diseased persons, or young children crying; to whose humours they are obliged to be subject, forced to bear with, and many times are poisoned with their nasty scents and crippled by the crowd of boxes and bundles? Is it for a man's health to travel with tired jades, to be laid fast in the foul ways and forced to wade up to the knees in mire; afterwards sit in the cold till teams of horses can be sent to pull the coach out? Is it for their health to travel in rotten coaches and to have their tackle, perch, or axle-tree broken, and then to wait three or four hours (sometimes half a day) to have them mended, and then to travel all night to make good their stage? Is it for a man's pleasure, or advantageous to his health and business, to travel with a mixed company that he knows not how to converse with; to be affronted by the rudeness of a surly, dogged, cursing, ill-natured coachman; necessitated to lodge or bait at the worst inn on the road, where there is no accommodation fit for gentlemen; and this merely because the owners of the inns and the coachmen are agreed together to cheat the guests?" Hence the writer loudly called for the immediate suppression of stage-coaches as a great nuisance and crying evil.

Travelling by coach was in early times a very deliberate affair. Time was of less consequence than safety, and coaches were advertised to start "God willing," and "about" such and such an hour "as shall seem good" to the majority of the passengers. The difference of a day in the journey from London to York was a small matter, and Thoresby was even accustomed to leave the coach and go in search of fossil shells in the fields on either side the road while making the journey between the two places. The long coach "put up" at sun-down, and "slept on the road." Whether the coach was to proceed or to stop at some favourite inn, was determined by the vote of the passengers, who usually appointed a chairman at the beginning of the journey.

In 1700, York was a week distant from London, and Tunbridge Wells, now reached in an hour, was two days. Salisbury and Oxford were also each a two days journey, Dover was three days, and Exeter five. The Fly coach from London to Exeter *slept* at the latter place the fifth night from town; the coach proceeding next morning to Axminster, where it breakfasted, and there a woman barber "*shaved* the coach." * Between London and Edinburgh, as late as 1763, a fortnight was consumed, the coach only starting once a month.† The risk of breaks-down in driving over the execrable roads may be inferred from the circumstance that every coach carried with it a box of carpenter's tools, and the hatchets were occasionally used in lopping off the branches of trees overhanging the road and obstructing the travellers' progress.

Some fastidious persons, disliking the slow travelling, as well as the promiscuous company which they ran the risk of encountering in the stage, were accustomed to advertise for partners in a postchaise, to share the charges and lessen the dangers of the road; and, indeed, to a sensitive person anything must have been preferable to the misery of travelling by the Canterbury stage, as thus described by a contemporary writer:—

> "On both sides squeez'd, how highly was I blest,
> Between two plump old women to be presst!
> A corp'ral fierce, a nurse, a child that cry'd,
> And a fat landlord, filled the other side.
> Scarce dawns the morning ere the cumbrous load
> Rolls roughly rumbling o'er the rugged road:
> One old wife coughs and wheezes in my ears,
> Loud scolds the other, and the soldier swears;
> Sour unconcocted breath escapes 'mine host,'
> The sick'ning child returns his milk and toast!"

* Roberts's 'Social History of the Southern Counties,' p. 494.—Little more than a century ago, we find the following advertisement of a Newcastle flying coach:—"May 9, 1734.—A coach will set out towards the end of next week for London, or any place on the road. To be performed in nine days,—being three days sooner than any other coach that travels the road; for which purpose eight stout horses are stationed at proper distances."

† In 1710 a Manchester manufacturer taking his family up to London, hired a coach for the whole way, which, in the then state of the roads, must have made it a journey of probably eight or ten days. And, in 1742, the system of travelling had so little improved, that a lady, wanting to come with her niece from Worcester to Manchester, wrote to a friend in the latter place to send her a hired coach, because the man *knew the road*, having brought from thence a family some time before."—Aikin's 'Manchester.'

When Samuel Johnson was taken by his mother to London in 1712, to have him touched by Queen Anne for "the evil," he relates,—"We went in the stage-coach and returned in the waggon, as my mother said, because my cough was violent; but the hope of saving a few shillings was no slight motive. . . . She sewed two guineas in her petticoat lest she should be robbed. . . . We were troublesome to the passengers; but to suffer such inconveniences in the stage-coach was common in those days to persons in much higher rank."

Mr. Pennant has left us the following account of his journey in the Chester stage to London in 1739-40: "The first day," says he, "with much labour, we got from Chester to Whitchurch, twenty miles; the second day to the 'Welsh Harp;' the third, to Coventry; the fourth, to Northampton; the fifth, to Dunstable; and, as a wondrous effort, on the last, to London, before the commencement of night. The strain and labour of six good horses, sometimes eight, drew us through the sloughs of Mireden and many other places. We were constantly out two hours before day, and as late at night, and in the depth of winter proportionally later. The single gentlemen, then a hardy race, equipped in jack-boots and trowsers, up to their middle, rode post through thick and thin, and, guarded against the mire, defied the frequent stumble and fall, arose and pursued their journey with alacrity; while, in these days, their enervated posterity sleep away their rapid journeys in easy chaises, fitted for the conveyance of the soft inhabitants of Sybaris."

No wonder, therefore, that a great deal of the travelling of the country continued to be performed on horseback, this being by far the pleasantest as well as most expeditious mode of journeying. On his marriage-day, Dr. Johnson rode from Birmingham to Derby with his Tetty, taking the opportunity of the journey to give his bride her first lesson in marital discipline. At a later period James Watt rode from Glasgow to London, when proceeding thither to learn the art of mathematical instrument making.

And it was a cheap and pleasant method of travelling when the weather was fine. The usual practice was, to buy a horse at the beginning of such a journey, and to sell the animal at the end of it. Dr. Skene, of Aberdeen, travelled from London to Edinburgh in 1753, being nineteen days on the road, the whole expenses of the journey amounting to only four guineas. The mare on which he rode, cost him eight guineas in London, and he sold her for the same price on his arrival in Edinburgh.

Nearly all the commercial gentlemen rode their own horses, carrying their samples and luggage in two bags at the saddle-bow; and hence their appellation of Riders or Bagmen. For safety's sake, they usually journeyed in company; for the dangers of travelling were not confined

THE NIGHT COACH. [After Rowlandson.]

merely to the ruggedness of the roads. The highways were infested by troops of robbers and vagabonds who lived by plunder. Turpin and Bradshaw beset the Great North Road; Duval, Macheath, Maclean, and hundreds of notorious highwaymen infested Hounslow Heath, Finchley Common, Shooter's Hill, and all the approaches to the metropolis. A very common sight then, was a gibbet erected by the roadside, with the skeleton of some malefactor hanging from it in chains; and "Hangman's-lanes" were especially numerous in the neighbourhood of London.* It was considered most unsafe to travel after dark, and when the first "night coach" was started, the risk was thought too great, and it was not patronised.

Travellers armed themselves on setting out upon a journey as if they were going to battle, and a blunderbuss was considered as indispensable for a coachman as a whip. Dorsetshire and Hampshire, like most other counties, were beset with gangs of highwaymen; and when the Grand Duke Cosmo set out from Dorchester to travel to London in 1669, he was "convoyed by a great many horse-soldiers belonging to the militia of the county, to secure him from robbers."† Thoresby, in his Diary, alludes with awe to his having passed safely "the great common where Sir Ralph Wharton slew the highwayman," and he also makes special mention of Stonegate Hole, "a notorious robbing place" near Grantham. Like every other traveller, that good man carried loaded pistols in his bags, and on one occasion he was thrown into great consternation near Topcliffe, in Yorkshire, on missing them, believing that

* Lord Campbell mentions the remarkable circumstance that Popham, afterwards Lord Chief Justice in the reign of Elizabeth, took to the road in early life, and robbed travellers on Gad's Hill. Highway robbery could not, however, have been considered a very ignominious pursuit at that time, as during Popham's youth a statute was made by which, on a first conviction for robbery, a peer of the realm or lord of parliament was entitled to have benefit of clergy, "though he cannot read!" What is still more extraordinary is, that Popham is supposed to have continued in his course as a highwayman even after he was called to the Bar. This seems to have been quite notorious, for when he was made Serjeant the wags reported that he served up some wine destined for an Alderman of London, which he had intercepted on its way from Southampton.—Aubrey, iii., 492.—Campbell's 'Chief Justices,' i., 210.

† 'Travels of Cosmo the Third, Grand Duke of Tuscany,' p. 147.

they had been abstracted by some designing rogues at the inn where he had last slept.* No wonder that, before setting out on a journey in those days, men were accustomed to make their wills.

When Mrs. Calderwood, of Coltness, travelled from Edinburgh to London in 1756, she relates in her Diary that she travelled in her own postchaise, attended by John Rattray, her stout serving man, on horseback, with pistols at his holsters, and a good broad sword by his side. The lady had also with her in the carriage a case of pistols, for use upon an emergency. Robberies were then of frequent occurrence in the neighbourhood of Bawtry, in Yorkshire; and one day a suspicious-looking character, whom they took to be a highwayman, made his appearance; but "John Rattray talking about powder and ball to the postboy, and showing his whanger, the fellow made off." Mrs. Calderwood started from Edinburgh on the 3rd of June, when the roads were dry and the weather was fine, and she reached London on the evening of the 10th, which was considered a rapid journey in those days.

The danger, however, from footpads and highwaymen was not greatest in remote country places, but in and about the metropolis itself. The proprietors of Bellsize House and gardens, in the Hampstead-road, then one of the principal places of amusement, had the way to London patrolled during the season by twelve "lusty fellows;" and Sadler's Wells, Vauxhall, and Ranelagh advertised

* "It is as common a custom, as a cunning policie in thieves, to place chamberlains in such great inns where cloathiers and graziers do lye; and by their large bribes to infect others, who were not of their own preferring; who noting your purses when you draw them, they'l gripe your cloak-bags, and feel the weight, and so inform the master thieves of what they think, and not those alone, but the Host himself is oft as base as they, if it be left in charge with them all night; he to his roaring guests either gives item, or shews the purse itself, who spend liberally, in hope of a speedie recruit." See 'A Brief yet Notable Discovery of Housebreakers,' &c., 1659. See also 'Street Robberies Considered; a Warning for Housekeepers,' 1676; 'Hanging not Punishment Enough,' 1701; &c.

similar advantages. Foot passengers proceeding towards Kensington and Paddington in the evening, would wait until a sufficiently numerous band had collected to set footpads at defiance, and then they started in company at known intervals, of which a bell gave due warning. Carriages were stopped in broad daylight in Hyde Park, and even in Piccadilly itself, and pistols presented at the breasts of fashionable people, who were called upon to deliver up their purses. Horace Walpole relates a number of curious instances of this sort, he himself having been robbed in broad day, with Lord Eglinton, Sir Thomas Robinson, Lady Albemarle, and many more. A curious robbery of the Portsmouth mail, in 1757, illustrates the imperfect postal communication of the period. The boy who carried the post had dismounted at Hammersmith, about three miles from Hyde Park Corner, and called for beer, when some thieves took the opportunity of cutting the mail-bag from off the horse's crupper and got away undiscovered!

The means adopted for the transport of merchandise were as tedious and difficult as those ordinarily employed for the conveyance of passengers. Corn and wool were sent to market on horses' backs,* manure was carried to the fields in panniers, and fuel was conveyed from the moss or the forest in the same way. During the winter months, the markets were inaccessible; and while in some localities the supplies of food were distressingly deficient, in others the superabundance actually rotted from the

* The food of London was then principally brought to town in panniers. The population being comparatively small, the feeding of London was still practicable in this way; besides, the city always possessed the great advantage of the Thames, which secured a supply of food by sea. In 'The Grand Concern of England Explained,' it is stated that the hay, straw, beans, peas, and oats, used in London, were principally raised within a circuit of twenty miles of the metropolis; but large quantities were also brought from Henley-on-Thames and other western parts, as well as from below Gravesend, by water; and many ships laden with beans came from Hull, and with oats from Lynn and Boston.

impossibility of consuming it or of transporting it to places where it was needed. The little coal used in the southern counties was principally sea-borne, though pack-horses occasionally carried coal inland for the supply of the black-

THE PACK-HORSE CONVOY. [By Louis Huard.]

smiths' forges. When Wollaton Hall was built by John of Padua for Sir Francis Willoughby in 1580, the stone was all brought on horses' backs from Ancaster, in Lincolnshire, thirty-five miles distant, and they loaded back with coal, which was taken in exchange for the stone.

The little trade which existed between one part of the kingdom and another was carried on by means of pack-horses, along roads little better than bridle-paths. These horses travelled in lines, with the bales or panniers strapped across their backs. The foremost horse bore a bell or a collar of bells, and was hence called the "bell-horse." He was selected because of his sagacity; and by the tinkling of the bells he carried, the movements of his followers were regulated. The bells also gave notice of the approach of the convoy to those who might be advancing from the opposite direction. This was a matter of some importance, as in many parts of the path there was not room for two loaded horses to pass each other, and quarrels and fights between the drivers of the pack-horse trains were frequent as to which of the meeting convoys was to pass down into the dirt and allow the other to pass along the bridleway. The pack-horses not only carried merchandise but passengers, and at certain times scholars proceeding to and from Oxford and Cambridge. When Smollett went from Glasgow to London, he travelled partly on pack-horse, partly by waggon, and partly on foot; and the adventures which he described as having befallen Roderick Random are supposed to have been drawn in a great measure from his own experiences during the journey.

A cross-country merchandise traffic gradually sprang up between the northern counties, since become pre-eminently the manufacturing districts of England; and long lines of pack-horses laden with bales of wool and cotton traversed the hill ranges which divide Yorkshire from Lancashire. Whitaker says that as late as 1753 the roads near Leeds consisted of a narrow hollow way little wider than a ditch, barely allowing of the passage of a vehicle drawn in a single line; this deep narrow road being flanked by an elevated causeway covered with flags or boulder stones. When travellers encountered each other on this narrow track, they often tried to wear out each other's patience rather than descend into the dirt alongside. The raw wool and bale goods of the district were nearly all carried along these flagged ways on the backs of single horses; and it is difficult to imagine the delay, the toil, and the perils by which the conduct of the traffic was attended. On horseback before daybreak and long after nightfall, these hardy sons of trade pursued their object with the spirit and intrepidity of foxhunters;

and the boldest of their country neighbours had no reason to despise either their horsemanship or their courage.* The Manchester trade was carried on in the same way. The chapmen used to keep gangs of pack-horses, which accompanied them to all the principal towns, bearing their goods in packs, which they sold to their customers, bringing back sheep's wool and other raw materials of manufacture.

The only records of this long-superseded mode of communication are now to be traced on the signboards of wayside public-houses. Many of the old roads still exist in Yorkshire and Lancashire; but all that remains of the former traffic is the pack-horse still painted on village sign-boards—things as retentive of odd bygone facts as the picture-writing of the ancient Mexicans.†

* 'Loides and Elmete,' by T. D. Whitaker, LL.D., 1816, p. 81. Notwithstanding its dangers, Dr. Whitaker seems to have been of opinion that the old mode of travelling was even safer than that which immediately followed it; "Under the old state of roads and manners," he says, "it was impossible that more than one death could happen at once; what, by any possibility, could take place analogous to a race betwixt two stage-coaches, in which the lives of thirty or forty distressed and helpless individuals are at the mercy of two intoxicated brutes?"

† In the curious collection of old coins at the Guildhall there are several halfpenny tokens issued by the proprietors of inns bearing the sign of the pack-horse. Some of these would indicate that pack-horses were kept for hire. We append a couple of illustrations of these curious old coins.

PACK-HORSE HALFPENNY TOKENS. [From the Guildhall Collection.]

CHAPTER III.

MANNERS AND CUSTOMS INFLUENCED BY THE STATE OF THE ROADS.

WHILE the road communications of the country remained thus imperfect, the people of one part of England knew next to nothing of the other. When a shower of rain had the effect of rendering the highways impassable, even horsemen were cautious in venturing far from home. But only a very limited number of persons could then afford to travel on horseback. The labouring people journeyed on foot, while the middle class used the waggon or the coach. But the amount of intercourse between the people of different districts—then exceedingly limited at all times—was, in a country so wet as England, necessarily suspended for all classes during the greater part of the year.

The imperfect communication existing between districts had the effect of perpetuating numerous local dialects, local prejudices, and local customs, which survive to a certain extent to this day; though they are rapidly disappearing, to the regret of many, under the influence of improved facilities for travelling. Every village had its witches, sometimes of different sorts, and there was scarcely an old house but had its white lady or moaning old man with a long beard. There were ghosts in the fens which walked on stilts, while the sprites of the hill country rode on flashes of fire. But the village witches and local ghosts have long since disappeared, excepting perhaps in a few of the less penetrable districts, where they may still survive.

It is curious to find that down even to the beginning of

the seventeenth century, the inhabitants of the southern districts of the island regarded those of the north as a kind of ogres. Lancashire was supposed to be almost impenetrable—as indeed it was to a considerable extent,—and inhabited by a half-savage race. Camden vaguely described it, previous to his visit in 1607, as that part of the country "lying beyond the mountains towards the Western Ocean." He acknowledged that he approached the Lancashire people "with a kind of dread," but determined at length "to run the hazard of the attempt," trusting in the Divine assistance. Camden was exposed to still greater risks in his survey of Cumberland. When he went into that county for the purpose of exploring the remains of antiquity it contained for the purposes of his great work, he travelled along the line of the Roman Wall as far as Thirlwall castle, near Haltwhistle; but there the limits of civilization and security ended; for such was the wildness of the country and of its lawless inhabitants beyond, that he was obliged to desist from his pilgrimage, and leave the most important and interesting objects of his journey unexplored.

About a century later, in 1700, the Rev. Mr. Brome, rector of Cheriton in Kent, entered upon a series of travels in England as if it had been a newly-discovered country. He set out in spring so soon as the roads had become passable. His friends convoyed him on the first stage of his journey, and left him, commending him to the Divine protection. He was, however, careful to employ guides to conduct him from one place to another, and in the course of his three years' travels he saw many new and wonderful things. He was under the necessity of suspending his travels when the winter or wet weather set in, and to lay up, like an arctic voyager, for several months, until the spring came round again. Mr. Brome passed through Northumberland into Scotland, then down the western side of the island towards Devonshire, where he found the farmers gathering in their corn on horse-back, the roads being so narrow that it was impossible for them to use waggons. He desired to travel into Cornwall, the boundaries of which he reached, but was prevented proceeding farther by the rains, and accordingly he made the best of his way home.*

The vicar of Cheriton was considered a wonderful man in his day,—almost as adventurous as we should now regard a traveller in Arabia. Twenty miles of slough, or an unbridged river between two parishes, were greater impediments to intercourse than the Atlantic Ocean now is between England and America. Considerable towns situated in the same county, were then more widely separated, for practical purposes, than London and Glasgow are at the present day. There were many districts which travellers never visited, and where the appearance of a stranger produced as great an excitement as the arrival of a white man in an African village.†

The author of 'Adam Bede' has given us a poet's picture of the leisure of last century, which has "gone where the spinning wheels are gone, and the pack-horses, and the slow waggons, and the pedlars who brought bargains to the door on sunny afternoons." Old Leisure "lived chiefly in the country, among pleasant seats and home-

* 'Three Years' Travels in England, Scotland, and Wales.' By James Brome, M.A., Rector of Cheriton, Kent. London, 1726.

† The treatment the stranger received was often very rude. When William Hutton, of Birmingham, accompanied by another gentleman, went to view the field of Bosworth, in 1770, "the inhabitants," he says, "set their dogs at us in the street, merely because we were strangers. Human figures not their own are seldom seen in these inhospitable regions. Surrounded with impassable roads, no intercourse with man to humanise the mind, nor commerce to smooth their rugged manners, they continue the boors of Nature." In certain villages in Lancashire and Yorkshire, not very remote from large towns, the appearance of a stranger, down to a comparatively recent period, excited a similar commotion amongst the villagers, and the word would pass from door to door, "Dost knaw 'im?" "Naya." "Is 'e straunger?" "Ey, for sewer." "Then paus' 'im—'Eave a duck [stone] at 'im—Fettle 'im!" And the "straunger" would straightway find the "ducks" flying about his head, and be glad to make his escape from the village with his life.

steads, and was fond of sauntering by the fruit-tree walls, and scenting the apricots when they were warmed by the morning sunshine, or sheltering himself under the orchard boughs at noon, when the summer pears were falling." But this picture has also its obverse side. Whole generations then lived a monotonous, ignorant, prejudiced, and humdrum life. They had no enterprize, no energy, little industry, and were content to die where they were born. The seclusion in which they were compelled to live, produced a picturesqueness of manners which is pleasant to look back upon, now that it is a thing of the past; but it was also accompanied with a degree of grossness and brutality much less pleasant to regard, and of which the occasional popular amusements of bull-running, cock-fighting, cock-throwing, the saturnalia of Plough-Monday, and such like, were the fitting exponents.

People then knew little except of their own narrow district. The world beyond was as good as closed against them. Almost the only intelligence of general affairs which reached them was communicated by pedlars and packmen, who were accustomed to retail to their customers the news of the day with their wares; or, at most, a news-letter from London, after it had been read nearly to pieces at the great house of the district, would find its way to the village, and its driblets of information would thus become diffused among the little community. Matters of public interest were long in becoming known in the remoter districts of the country. Macaulay relates that the death of Queen Elizabeth was not heard of in some parts of Devon until the courtiers of her successor had ceased to wear mourning for her. The news of Cromwell's being made Protector only reached Bridgewater nineteen days after the event, when the bells were set a-ringing; and the churches in the Orkneys continued to put up the usual prayers for James II. three months after he had taken up his abode at St. Germains.

There were then no shops in the smaller towns or villages, and comparatively few in the larger; and these were badly furnished with articles for general use. The country people were irregularly supplied by hawkers, who sometimes bore their whole stock upon their back, or occasionally on that of their pack-horses. Pots, pans, and household utensils were sold from door to door. Until a comparatively recent period, the whole of the pottery-ware manufactured in Staffordshire was hawked about and disposed of in this way. The pedlars carried frames resembling camp-stools, on which they were accustomed to display their wares when the opportunity occurred for showing them to advantage. The articles which they sold were chiefly of a fanciful kind—ribbons, laces, and female finery; the housewives' great reliance for the supply of general clothing in those days being on domestic industry.

Every autumn, the mistress of the household was accustomed to lay in a store of articles sufficient to serve for the entire winter. It was like laying in a stock of provisions and clothing for a siege during the time that the roads were closed. The greater part of the meat required for winter's use was killed and salted down at Martinmas, while stockfish and baconed herrings were provided for Lent. Scatcherd says that in his district the clothiers united in groups of three or four, and at the Leeds winter fair they would purchase an ox, which, having divided, they salted and hung the pieces for their winter's food.* There was also the winter's stock of firewood to be provided, and the rushes with which to strew the floors—carpets being a comparatively modern invention; besides, there was the store of wheat and barley for bread, the malt for ale, the honey for sweetening (then used for sugar), the salt, the spiceries, and the savoury herbs so much employed in the ancient cookery. When the stores were laid in, the housewife was in a position to bid defiance to bad roads for six months to come. This was the case of the well-to-do; but the poorer

* Scatcherd, 'History of Morley.'

classes, who could not lay in a store for winter, were often very badly off both for food and firing, and in many hard seasons they literally starved. But charity was active in those days, and many a poor man's store was eked out by his wealthier neighbour.

When the household supply was thus laid in, the mistress, with her daughters and servants, sat down to their distaffs and spinning-wheels; for the manufacture of the family clothing was usually the work of the winter months. The fabrics then worn were almost entirely of wool, silk and cotton being scarcely known. The wool, when not grown on the farm, was purchased in a raw state, and was carded, spun, dyed, and in many cases woven at home: so also with the linen clothing, which, until quite a recent date, was entirely the produce of female fingers and household spinning-wheels. This kind of work occupied the winter months, occasionally alternated with knitting, embroidery, and tapestry work. Many of our country houses continue to bear witness to the steady industry of the ladies of even the highest ranks in those times, in the fine tapestry hangings with which the walls of many of the older rooms in such mansions are covered.

Among the humbler classes, the same winter's work went on. The women sat round log fires knitting, plaiting, and spinning by fire-light, even in the daytime. Glass had not yet come into general use, and the openings in the wall which in summer-time served for windows, had necessarily to be shut close with boards to keep out the cold, though at the same time they shut out the light. The chimney, usually of lath and plaster, ending overhead in a cone and funnel for the smoke, was so roomy in old cottages as to accommodate almost the whole family sitting around the fire of logs piled in the reredosse in the middle, and there they carried on their winter's work.

Such was the domestic occupation of women in the rural districts in olden times; and it may perhaps be questioned whether the revolution in our social system, which has taken out of their hands so many branches of household manufacture and useful domestic employment, be an altogether unmixed blessing.

Winter at an end, and the roads once more available for travelling, the Fair of the locality was looked forward to with interest. Fairs were among the most important institutions of past times, and were rendered necessary by the imperfect road communications. The right of holding them was regarded as a valuable privilege, conceded by the sovereign to the lords of the manors, who adopted all manner of devices to draw crowds to their markets. They were usually held at the entrances to valleys closed against locomotion during winter, or in the middle of rich grazing districts, or, more frequently, in the neighbourhood of famous cathedrals or churches frequented by flocks of pilgrims. The devotion of the people being turned to account, many of the fairs were held on Sundays in the churchyards; and almost in every parish a market was instituted on the day on which the parishioners were called together to do honour to their patron saint.

The local fair, which was usually held at the beginning or end of winter, often at both times, became the great festival as well as market of the district; and the business as well as the gaiety of the neighbourhood usually centred on such occasions. High courts were held by the Bishop or Lord of the Manor, to accommodate which special buildings were erected, used only at fair time. Among the fairs of the first class in England were Winchester, St. Botolph's Town (Boston), and St. Ives. We find the great London merchants travelling thither in caravans, bearing with them all manner of goods, and bringing back the wool purchased by them in exchange.

Winchester Great Fair attracted merchants from all parts of Europe. It was held on the hill of St. Giles, and was divided into streets of booths, named after the merchants of the different countries who exposed their wares in them. "The passes through the great woody districts, which

English merchants coming from London and the West would be compelled to traverse, were on this occasion carefully guarded by mounted 'serjeants-at-arms,' since the wealth which was being conveyed to St. Giles's-hill attracted bands of outlaws from all parts of the country." * Weyhill Fair, near Andover, was another of the great fairs in the same district, which was to the West country agriculturists and clothiers what Winchester St. Giles's Fair was to the general merchants.

The principal fair in the northern districts was that of St. Botolph's Town (Boston), which was resorted to by people from great distances to buy and sell commodities of various kinds. Thus we find, from the 'Compotus' of Bolton Priory,† that the monks of that house sent their wool to St. Botolph's Fair to be sold, though it was a good hundred miles distant; buying in return their winter supply of groceries, spiceries, and other necessary articles. That fair, too, was often beset by robbers, and on one occasion a strong party of them, under the disguise of monks, attacked and robbed certain booths, setting fire to the rest; and such was the amount of destroyed wealth, that it is said the veins of molten gold and silver ran along the streets.

The concourse of persons attending these fairs was immense. The nobility and gentry, the heads of the religious houses, the yeomanry and the commons, resorted to them to buy and sell all manner of agricultural produce. The farmers there sold their wool and cattle, and hired their servants; while their wives disposed of the surplus produce of their winter's industry, and bought their cutlery, bijouterie, and more tasteful articles of apparel. There were caterers there for all customers; and stuffs and wares were offered for sale from all countries. And in the wake of this business part of the fair there invariably followed a crowd of ministers to the popular tastes—quack doctors and merry andrews,

* Murray's 'Handbook of Surrey, Hants, and Isle of Wight,' 168.
† Whitaker's 'History of Craven.'

juggiers and minstrels, singlestick players, grinners through horse-collars, and sportmakers of every kind.

Smaller fairs were held in most districts for similar purposes of exchange. At these the staples of the locality were sold and servants usually hired. Many were for special purposes—cattle fairs, leather fairs, cloth fairs, bonnet fairs, fruit fairs. Scatcherd says that less than a century ago a large fair was held between Huddersfield and Leeds, in a

SITE OF AN ANCIENT BRITISH VILLAGE AND FAIR ON DARTMOOR.
[By Percival Skelton.]

field still called Fairstead, near Birstal, which used to be a great mart for fruit, onions, and such like; and that the clothiers resorted thither from all the country round to purchase the articles, which were stowed away in barns, and sold at booths by lamplight in the morning.*

Even Dartmoor had its fair, on the site of an ancient

* Scatcherd's 'History of Morley,' 226.

British village or temple near Merivale Bridge, testifying to its great antiquity; for it is surprising how an ancient fair lingers about the place on which it has been accustomed to be held, long after the necessity for it has ceased. The site of this old fair at Merivale Bridge is the more curious, as in its immediate neighbourhood, on the road between Two Bridges and Tavistock, is found the singular-looking granite rock, bearing so remarkable a resemblance to the Egyptian sphynx, in a mutilated state. It is of similarly colossal proportions, and stands in a district almost as lonely as that in which the Egyptian sphynx looks forth over the sands of the Memphean Desert.*

The last occasion on which the fair was held in this secluded spot was in the year 1625, when the plague raged at Tavistock; and there is a part of the ground, situated amidst a line of pillars marking a stone avenue—a characteristic feature of the ancient aboriginal worship—which is to this day pointed out and called by the name of the "Potatoe market."

But the glory of the great fairs has long since departed. They declined with the extension of turnpikes, and railroads gave them their death-blow. Shops now exist in every little town and village, drawing their supplies regularly by road and canal from the most distant parts. St. Bartholomew, the great fair of London,† and Donnybrook, the great fair

* Vixen Tor is the name of this singular-looking rock. But it is proper to add, that its appearance is probably accidental, the head of the Sphynx being produced by the three angular blocks of rock seen in profile. Mr. Borlase, however, in his 'Antiquities of Cornwall,' expresses the opinion that the rock-basins on the summit of the rock were used by the Druids for purposes connected with their religious ceremonies.

† The provisioning of London, now grown so populous, would be almost impossible but for the perfect system of roads now converging on it from all parts. In early times, London, like country places, had to lay in its stock of salt-provisions against winter, drawing its supplies of vegetables from the country within easy reach of the capital. Hence the London market-gardeners petitioned against the extension of turnpike-roads about a century ago, as they afterwards petitioned against the extension of railways, fearing lest their trade should be destroyed by the competition of country-grown cabbages. But the extension of the roads had

of Dublin, have been suppressed as nuisances; and nearly all that remains of the dead but long potent institution of the Fair, is the occasional exhibition at periodic times in country places, of pig-faced ladies, dwarfs, giants, double-bodied calves, and such-like wonders, amidst a blatant clangour of drums, gongs, and cymbals. Like the sign of the Pack-Horse over the village inn door, the modern village fair, of which the principal article of merchandise is gingerbread-nuts, is but the vestige of a state of things that has long since passed away.

There were, however, remote and almost impenetrable districts which long resisted modern inroads. Of such was Dartmoor, which we have already more than once referred to. The difficulties of road-engineering in that quarter, as well as the sterility of a large proportion of the moor, had the effect of preventing its becoming opened up to modern

---

become a matter of absolute necessity, in order to feed the huge and ever-increasing mouth of the Great Metropolis, the population of which has grown in about two centuries from four hundred thousand to three millions. This enormous population has, perhaps, never at any time more than a fortnight's supply of food in stock, and most families not more than a few days; yet no one ever entertains the slightest apprehension of a failure in the supply, or even of a variation in the price from day to day in consequence of any possible shortcoming. That this should be so, would be one of the most surprising things in the history of modern London, but that it is sufficiently accounted for by the magnificent system of roads, canals, and railways, which connect it with the remotest corners of the kingdom. Modern London is mainly fed by steam. The Express Meat-Train, which runs nightly from Aberdeen to London, drawn by two engines, and makes the journey in twenty-four hours, is but a single illustration of the rapid and certain method by which modern London is fed. The north Highlands of Scotland have thus, by means of railways, become grazing-grounds for the metropolis. Express fish-trains from Dunbar and Eyemouth (Smeaton's harbours), augmented by fish-trucks from Cullercoats and Tynemouth on the Northumberland coast, and from Redcar, Whitby, and Scarborough on the Yorkshire coast, also arrive in London every morning. And what with steam-vessels bearing cattle, and meat and fish arriving by sea, and canal-boats laden with potatoes from inland, and railway-vans laden with butter and milk drawn from a wide circuit of country, and road-vans piled high with vegetables within easy drive of Covent Garden, the Great Mouth is thus from day to day regularly, satisfactorily, and expeditiously filled.

traffic; and it is accordingly curious to find how much of its old manners, customs, traditions, and language has been preserved. It looks like a piece of England of the Middle Ages, left behind on the march. Witches still hold their sway on Dartmoor, where there exist no less than three distinct kinds—white, black, and grey,*—and there are still professors of witchcraft, male as well as female, in most of the villages.

As might be expected, the pack-horses held their ground in Dartmoor the longest, and in some parts of North Devon they are not yet extinct. When our artist was in the neighbourhood, sketching the ancient bridge on the moor and the site of the old fair, a farmer said to him, "I well remember the train of pack-horses and the effect of their jingling bells on the silence of Dartmoor. My grandfather, a respectable farmer in the north of Devon, was the first to use a 'butt' (a square box without wheels, dragged by a horse) to carry manure to field; he was also the first man in the district to use an umbrella, which on Sundays he hung in the church-porch, an object of curiosity to the villagers." We are also informed by a gentleman who resided for some time at South Brent, on the borders of the Moor, that the introduction of the first cart in that district is remembered by many now living, the bridges having been shortly afterwards widened to accommodate the wheeled vehicles.

The primitive features of this secluded district are perhaps best represented by the interesting little town of Chagford, situated in the valley of the North Teign, an ancient stannary and market town backed by a wide stretch of moor. The houses of the place are built of moor stone—grey, venerable-looking, and substantial—some with projecting porch and parvise room over, and granite-mullioned windows; the ancient church, built of granite, with a stout old steeple of the same material, its embattled porch and

* The white witches are kindly disposed, the black cast the "evil eye," and the grey are consulted for the discovery of theft, &c.

granite-groined vault springing from low columns with Norman-looking capitals, forming the sturdy centre of this ancient town clump.

A post-chaise is still a phenomenon in Chagford, the roads and lanes leading to it being so steep and rugged as to be ill adapted for springed vehicles of any sort. The upland road or track to Tavistock scales an almost precipitous hill, and though well enough adapted for the pack-horse of the last century, it is quite unfitted for the cart and waggon traffic of this. Hence the horse with panniers maintains its ground in the Chagford district; and the double-horse, furnished with a pillion for the lady riding behind, is still to be met with in the country roads.

Among the patriarchs of the hills, the straight-breasted blue coat may yet be seen, with the shoe fastened with buckle and strap as in the days when George III. was king; and old women are still found retaining the cloak and hood of their youth. Old agricultural implements continue in use. The slide or sledge is seen in the fields; the flail, with its monotonous strokes, resounds from the barn-floors; the corn is sifted by the windstow—the wind merely blowing away the chaff from the grain when shaken out of sieves by the motion of the hand on some elevated spot; the old wooden plough is still at work, and the goad is still used to urge the yoke of oxen in dragging it along.

"In such a place as Chagford," says Mr. Rowe, "the cooper or rough carpenter will still find a demand for the pack-saddle, with its accompanying furniture of *crooks*, *crubs*, or *dung-pots*. Before the general introduction of carts, these rough and ready contrivances were found of great utility in the various operations of husbandry, and still prove exceedingly convenient in situations almost, or altogether, inaccessible to wheel-carriages. The *long crooks* are used for the carriage of corn in sheaf from the harvest-field to the mowstead or barn, for the removal of furze, browse, faggot-wood, and other light materials. The writer of one of the happiest effusions of the local muse,* with fidelity to nature equal to Cowper or Crabbe, has

* See 'The Devonshire Lane,' above quoted, note to p. 7.

introduced the figure of a Devonshire pack-horse bending under the 'swagging load' of the high-piled *crooks* as an emblem of care toiling along the narrow and rugged path of life. The force and point of the imagery must be lost to those who have never seen (and, as in an instance which came under my own knowledge, never heard

THE DEVONSHIRE CROOKS.
[By Louis Huard, after an Original Sketch.]

of) this unique specimen of provincial agricultural machinery. The crooks are formed of two poles,* about ten feet long, bent, when green, into the required curve, and when dried in that shape are

* Willow saplings, crooked and dried in the required form.

connected by horizontal bars. A pair of crooks, thus completed, is slung over the pack-saddle—one 'swinging on each side to make the balance true.' The short crooks, or *crubs*, are slung in a similar manner. These are of stouter fabric, and angular shape, and are used for carrying logs of wood and other heavy materials. The dung-pots, as the name implies, were also much in use in past times, for the removal of dung and other manure from the farm-yard to the fallow or plough lands. The *slide*, or sledge, may also still occasionally be seen in the hay or corn fields, sometimes without, and in other cases mounted on low wheels, rudely but substantially formed of thick plank, such as might have brought the ancient Roman's harvest load to the barn some twenty centuries ago."

Mrs. Bray says the crooks are called by the country people " Devil's tooth-picks." A correspondent informs us that the queer old crook-packs represented in our illustration are still in use in North Devon. He adds: " The pack-horses were so accustomed to their position when travelling in line (going in double file) and so jealous of their respective places, that if one got wrong and took another's place, the animal interfered with would strike at the offender with his crooks."

CHAPTER IV.

ROADS AND TRAVELLING IN SCOTLAND IN THE LAST CENTURY.

THE internal communications of Scotland, which Telford did so much in the course of his life to improve, were, if possible, even worse than those of England about the middle of last century. The land was more sterile, and the people were much poorer. Indeed, nothing could be more dreary than the aspect which Scotland then presented. Her fields lay untilled, her mines unexplored, and her fisheries uncultivated. The Scotch towns were for the most part collections of thatched mud cottages, giving scant shelter to a miserable population. The whole country was desponding, gaunt, and haggard, like Ireland in its worst times. The common people were badly fed and wretchedly clothed, those in the country for the most part living in huts with their cattle. Lord Kaimes said of the Scotch tenantry of the early part of last century, that they were so benumbed by oppression and poverty that the most able instructors in husbandry could have made nothing of them. A writer in the 'Farmer's Magazine' sums up his account of Scotland at that time in these words:—"Except in a few instances, it was little better than a barren waste." *

The modern traveller through the Lothians—which now exhibit perhaps the finest agriculture in the world—will scarcely believe that less than a century ago these counties were mostly in the state in which Nature had left them. In the interior there was little to be seen but bleak moors and quaking bogs. The chief part of each farm consisted

* 'Farmer's Magazine,' 1803. No. xiii. p. 101.

of "out-field," or unenclosed land, no better than moorland, from which the hardy black cattle could scarcely gather herbage enough in winter to keep them from starving. The "in-field" was an enclosed patch of ill-cultivated ground, on which oats and "bear," or barley, were grown; but the principal crop was weeds.

Of the small quantity of corn raised in the country, nine-tenths were grown within five miles of the coast; and of wheat very little was raised—not a blade north of the Lothians. When the first crop of that grain was tried on a field near Edinburgh, about the middle of last century, people flocked to it as a wonder. Clover, turnips, and potatoes had not yet been introduced, and no cattle were fattened: it was with difficulty they could be kept alive.

All loads were as yet carried on horseback; but when the farm was too small, or the crofter too poor to keep a horse, his own or his wife's back bore the load. The horse brought peats from the bog, carried the oats or barley to market, and bore the manure a-field. But the uses of manure were as yet so little understood that, if a stream were near, it was usually thrown in and floated away, and in summer it was burnt.

What will scarcely be credited, now that the industry of Scotland has become educated by a century's discipline of work, was the inconceivable listlessness and idleness of the people. They left the bog unreclaimed, and the swamp undrained. They would not be at the trouble to enclose lands easily capable of cultivation. There was, perhaps, but little inducement on the part of the agricultural class to be industrious; for they were too liable to be robbed by those who preferred to be idle. Andrew Fletcher, of Saltoun—commonly known as "The Patriot," because he was so strongly opposed to the union of Scotland with England*—published a pamphlet, in 1698, strikingly illus-

* Bad although the condition of Scotland was at the beginning of last century, there were many who believed that it would be made

trative of the lawless and uncivilized state of the country at that time. After giving a dreadful picture of the then state of Scotland: two hundred thousand vagabonds begging from door to door and robbing and plundering the poor people,—"in years of plenty many thousands of them meeting together in the mountains, where they feast and riot for many days; and at country weddings, markets, burials, and other like public occasions, they are to be seen, both men and women, perpetually drunk, cursing, blaspheming, and fighting together,"—he proceeded to urge that every man of a certain estate should be obliged to take a proportionate number of these vagabonds and compel them to work for him; and further, that such serfs, with their wives and children, should be incapable of alienating their service from their master or owner until he had been reimbursed for the money he had expended on them: in other words, their owner was to have the power of selling them. "The Patriot" was, however, aware that "great address, diligence, and severity" were required to carry out his scheme; "for," said he, "that sort of people are so desperately wicked, such enemies of all work and labour, and, which is yet more amazing, so proud in esteeming their own condition above that which they will be sure to call Slavery, that unless prevented by the utmost industry and diligence, upon the first publication of any orders necessary for putting in execution such a design, they will rather die with hunger in caves and dens, and murder

---

worse by the carrying of the Act of Union. The Earl of Wigton was one of these. Possessing large estates in the county of Stirling, and desirous of taking every precaution against what he supposed to be impending ruin, he made over to his tenants, on condition that they continued to pay him their then low rents, his extensive estates in the parishes of Denny, Kirkintulloch, and Cumbernauld, retaining only a few fields round the family mansion ['Farmer's Magazine,' 1808, No. xxxiv. p. 193]. Fletcher of Saltoun also feared the ruinous results of the Union, though he was less precipitate in his conduct than the Earl of Wigton. We need scarcely say how entirely such apprehensions were falsified by the actual results.

their young children, than appear abroad to have them and themselves taken into such service." *

Although the recommendations of Andrew Fletcher of Saltoun were embodied in no Act of Parliament, the magistrates of some of the larger towns did not hesitate to kidnap and sell into slavery lads and men found lurking in the streets, which they continued to do down to a comparatively recent period. This, however, was not so surprising as that at the time of which we are speaking, and, indeed, until the end of last century, there was a veritable slave class in Scotland—the class of colliers and salters—who were bought and sold with the estates to which they belonged, as forming part of the stock. When they ran away, they were advertised for, as negroes were in the American States until within the last few years. It is curious, in turning over an old volume of the 'Scots Magazine,' to find a General Assembly's petition to Parliament for the abolition of slavery in America almost alongside the report of a trial of some colliers who had absconded from a mine near Stirling to which they belonged. But the degraded condition of the home slaves then excited comparatively little interest. Indeed, it was not until the very last year of the last century that prædial slavery was abolished in Scotland—only three short reigns ago, almost within the memory of men still living.†

* 'Fletcher's Political Works,' London, 1737, p. 149. As the population of Scotland was then only about 1,200,000, the beggars of the country, according to the above account, must have constituted about one-sixth of the whole community.

† Act 39th George III. c. 56. See 'Lord Cockburn's Memorials,' pp. 76-9. As not many persons may be aware how recent has been the abolition of slavery in Britain, the author of this book may mention the fact, that he personally knew a man who had been "born a slave in Scotland," to use his own words, and lived to tell it. He had resisted being transferred to another owner on the sale of the estate to which he was "bound," and refused to "go below," on which he was imprisoned in Edinburgh gaol, where he lay for a considerable time. The case excited much interest, and probably had some effect in leading to the alteration in the law relating to colliers and salters which shortly after followed.

The greatest resistance was offered to the introduction of improvements in agriculture, though it was only at rare intervals that these were attempted. There was no class possessed of enterprise or wealth. An idea of the general poverty of the country may be inferred from the fact that about the middle of last century the whole circulating medium of the two Edinburgh banks—the only institutions of the kind then in Scotland—amounted to only 200,000*l.*, which was sufficient for the purposes of trade, commerce, and industry. Money was then so scarce that Adam Smith says it was not uncommon for workmen, in certain parts of Scotland, to carry nails instead of pence to the baker's or the alehouse. A middle class could scarcely as yet be said to exist, or any condition between the starving cottiers and the impoverished proprietors, whose available means were principally expended in hard drinking.* The latter were, for the most part, too proud and too ignorant to interest themselves in the improvement of their estates; and the few who did so had very little encouragement to persevere. Miss Craig, in describing the efforts made by her father, William Craig, laird of Arbigland, in Kirkcudbright, says, "The indolent obstinacy of the lower class of the people was found to be almost unconquerable. Amongst other instances of their laziness, I have heard him say that, upon the introduction of the mode of dressing the grain at night which had been thrashed during the day, all the servants in the neighbourhood refused to adopt the measure, and even threatened to destroy the houses of their employers by fire if they continued to insist upon the business. My father speedily perceived that a forcible remedy was required for the evil. He gave his servants the choice of removing the thrashed grain in the evening, or becoming inhabitants of Kirkcudbright gaol: they preferred the former alternative, and open murmurings were no longer heard." †

---

* See 'Autobiography of Dr. Alexander Carlyle,' *passim.*

† 'Farmer's Magazine,' June, 1811, No. xlvi. p. 155.

The wages paid to the labouring classes were then very low. Even in East Lothian, which was probably in advance of the other Scotch counties, the ordinary day's wage of a labouring man was only five pence in winter and six pence in summer. Their food was wholly vegetable, and was insufficient in quantity as well as bad in quality. The little butcher's meat consumed by the better class was salted beef and mutton, stored up in Ladner time (between Michaelmas and Martinmas) for the year's consumption. Mr. Buchan Hepburn says the Sheriff of East Lothian informed him that he remembered when not a bullock was slaughtered in Haddington market for a whole year, except at that time; and, when Sir David Kinloch, of Gilmerton, sold ten wedders to an Edinburgh butcher, he stipulated for three several terms to take them away, to prevent the Edinburgh market from being overstocked with fresh butcher's meat!*

The rest of Scotland was in no better state: in some parts it was even worse. The rich and fertile county of Ayr, which now glories in the name of "the garden of Scotland," was for the most part a wild and dreary waste, with here and there a poor, miserable, comfortless hut, where the farmer and his family lodged. There were no enclosures of land, except one or two about a proprietor's residence; and black cattle roamed at large over the face of the country. When an attempt was made to enclose the lands for the purposes of agriculture, the fences were levelled by the dispossessed squatters. Famines were frequent among the poorer classes; the western counties not producing food enough for the sustenance of the inhabitants, few though they were in number. This was also the case in Dumfries, where the chief part of the grain required for the population was brought in "tumbling-cars" from the sandbeds of Esk; "and when the waters

† See Buchan Hepburn's 'General View of the Agriculture and Economy of East Lothian,' 1794, p. 95.

were high by reason of spates [or floods], and there being no bridges, so that the cars could not come with the meal, the tradesmen's wives might be seen in the streets of Dumfries, crying because there was no food to be had." *

The misery of the country was enormously aggravated by the wretched state of the roads. There were, indeed, scarcely any made roads throughout the country. Hence the communication between one town and another was always difficult, especially in winter. There were only rough tracks across moors, and when one track became too deep, another alongside of it was chosen, and was in its turn abandoned, until the whole became equally impassable. In wet weather these tracks became "mere sloughs, in which the carts or carriages had to slumper through in a half-swimming state, whilst in times of drought it was a continual jolting out of one hole into another." †

Such being the state of the highways, it will be obvious that very little communication could exist between one part of the country and another. Single-horse traffickers, called cadgers, plied between the country towns and the villages, supplying the inhabitants with salt, fish, earthenware, and articles of clothing, which they carried in sacks or creels hung across their horses' backs. Even the trade between Edinburgh and Glasgow was carried on in the same primitive way, the principal route being along the high grounds west of Boroughstoness, near which the remains of the old pack-horse road are still to be seen.

It was long before vehicles of any sort could be used on the Scotch roads. Rude sledges and tumbling-cars were employed near towns, and afterwards carts, the wheels of which were first made of boards. It was long before travelling by coach could be introduced in Scotland. When Smollett travelled from Glasgow to Edinburgh on his way to London, in 1739, there was neither coach, cart,

* Letter of John Maxwell, in Appendix to Macdiarmid's 'Picture of Dumfries,' 1823.

† Robertson's 'Rural Recollections,' p. 38.

nor waggon on the road. He accordingly accompanied the pack-horse carriers as far as Newcastle, "sitting upon a pack-saddle between two baskets, one of which," he says, "contained my goods in a knapsack."

In 1743 an attempt was made by the Town Council of Glasgow to set up a stage-coach or "lando." It was to be drawn by six horses, carry six passengers, and run between Glasgow and Edinburgh, a distance of forty-four miles, once a week in winter, and twice a week in summer. The project, however, seems to have been thought too bold for the time, for the "lando" was never started. It was not until the year 1749 that the first public conveyance, called "The Glasgow and Edinburgh Caravan," was started between the two cities, and it made the journey between the one place and the other in two days. Ten years later another vehicle was started, named "The Fly" because of its unusual speed, and it contrived to make the journey in rather less than a day and a half.

About the same time, a coach with four horses was started between Haddington and Edinburgh, and it took a full winter's day to perform the journey of sixteen miles: the effort being to reach Musselburgh in time for dinner, and go into town in the evening. As late as 1763 there was only one stage-coach in all Scotland in communication with London, and that set out from Edinburgh only once a month. The journey to London occupied from ten to fifteen days, according to the state of the weather; and those who undertook so dangerous a journey usually took the precaution of making their wills before starting.

When carriers' carts were established, the time occupied by them on the road will now appear almost incredible. Thus the common carrier between Selkirk and Edinburgh, a distance of only thirty-eight miles, took about a fortnight to perform the double journey. Part of the road lay along Gala Water, and in summer time, when the river-bed was dry, the carrier used it as a road. The townsmen of this adventurous individual, on the morning of his way-going, were accustomed to turn out and take leave of him, wishing him a safe return from his perilous journey. In winter the route was simply impracticable, and the communication was suspended until the return of dry weather.

While such was the state of the communications in the immediate neighbourhood of the metropolis of Scotland, matters were, if possible, still worse in the remoter parts of the country. Down to the middle of last century, there were no made roads of any kind in the south-western counties. The only inland trade was in black cattle; the tracks were impracticable for vehicles, of which there were only a few —carts and tumbling-cars—employed in the immediate neighbourhood of the towns. When the Marquis of Downshire attempted to make a journey through Galloway in his coach, about the year 1760, a party of labourers with tools attended him, to lift the vehicle out of the ruts and put on the wheels when it got dismounted. Even with this assistance, however, his Lordship occasionally stuck fast, and when within about three miles of the village of Creetown, near Wigton, he was obliged to send away the attendants, and pass the night in his coach on the Corse of Slakes with his family.

Matters were, of course, still worse in the Highlands, where the rugged character of the country offered formidable difficulties to the formation of practicable roads, and where none existed save those made through the rebel districts by General Wade shortly after the rebellion of 1715. The people were also more lawless and, if possible, more idle, than those of the Lowland districts about the same period. The latter regarded their northern neighbours as the settlers in America did the Red Indians round their borders—like so many savages always ready to burst in upon them, fire their buildings, and carry off their cattle.* Very little corn was grown in the neighbourhood

* Very little was known of the geography of the Highlands down to the beginning of the seventeenth century. The principal information

of the Highlands, on account of its being liable to be reaped and carried off by the caterans, and that before it was ripe. The only method by which security of a certain sort could be obtained was by the payment of blackmail to some of the principal chiefs, though this was not sufficient to protect them against the lesser marauders. Regular contracts were drawn up between proprietors in the counties of Perth, Stirling, and Dumbarton, and the Macgregors, in which it was stipulated that if less than seven cattle were stolen—which peccadillo was known as *picking*—no redress should be required; but if the number stolen exceeded seven—such amount of theft being raised to the dignity of *lifting*—then the Macgregors were bound to recover. This blackmail was regularly levied as far south as Campsie—then within six miles of Glasgow, but now almost forming part of it—down to within a few months of the outbreak of the Rebellion of 1745.*

Under such circumstances, agricultural improvement was altogether impossible. The most fertile tracts were allowed to lie waste, for men would not plough or sow where they had not the certain prospect of gathering in

on the subject being derived from Danish materials. It appears, however, that in 1608, one Timothy Pont, a young man without fortune or patronage, formed the singular resolution of travelling over the whole of Scotland, with the sole view of informing himself as to the geography of the country, and he persevered to the end of his task through every kind of difficulty; exploring all the islands with the zeal of a missionary, though often pillaged and stript of everything by the then barbarous inhabitants. The enterprising youth received no recognition nor reward for his exertions, and he died in obscurity, leaving his maps and papers to his heirs. Fortunately, James I. heard of the existence of Pont's papers, and purchased them for public use. They lay, however, unused for a long time in the offices of the Scotch Court of Chancery, until they were at length brought to light by Mr. Robert Gordon, of Straloch, who made them the basis of the first map of Scotland having any pretensions to accuracy that was ever published.

* Mr. Grant, of Corrymorry, used to relate that his father, when speaking of the Rebellion of 1745, always insisted that a rising in the Highlands was *absolutely necessary* to give employment to the numerous bands of lawless and idle young men who infested every property.—Anderson's 'Highlands and Islands of Scotland,' p. 432.

the crop. Another serious evil was, that the lawless habits of their neighbours tended to make the Lowland borderers almost as ferocious as the Higlanders themselves. Feuds were of constant occurrence between neighbouring baronies, and even contiguous parishes; and the country fairs, which were tacitly recognised as the occasions for settling quarrels, were the scenes of as bloody faction fights as were ever known in Ireland even in its worst days. When such was the state of Scotland only a century ago, what may we not hope for from Ireland when the civilizing influences of roads, schools, and industry have made more general progress amongst her people?

Yet Scotland had not always been in this miserable condition. There is good reason to believe that as early as the thirteenth century, agriculture was in a much more advanced state than we find it to have been in the eighteenth. It would appear from the extant chartularies of monastic establishments, which then existed all over the Lowlands, that a considerable portion of their revenue was derived from wheat, which also formed no inconsiderable part of their living. The remarkable fact is mentioned by Walter de Hemingford, the English historian, that when the castle of Dirleton, in East Lothian, was besieged by the army of Edward I., in the beginning of July, 1298, the men, being reduced to great extremities for provisions, were fain to subsist on the pease and beans which they gathered in the fields.* This statement is all the more remarkable on two accounts: first, that pease and beans should then have been so plentiful as to afford anything like sustenance for an army; and second, that they should have been fit for use so early in the season, even allowing for the difference between the old and new styles in the reckoning of time.

The magnificent old abbeys and churches of Scotland in early times also indicate that at some remote period

---

* 'Lord Hailes's Annals,' i., 379.

a degree of civilization and prosperity prevailed, from which the country had gradually fallen. The ruins of the ancient edifices of Melrose, Kilwinning, Aberborthwick, Elgin, and other religious establishments, show that architecture must then have made great progress in the North, and lead us to the conclusion that the other arts had reached a like stage of advancement. This is borne out by the fact of the number of well-designed and well-built bridges of olden times which still exist in different parts of Scotland. "And when we consider," says Professor Innes, "the long and united efforts required in the early state of the arts for throwing a bridge over any considerable river, the early occurrence of bridges may well be admitted as one of the best tests of civilization and national prosperity." * As in England, so in Scotland, the reclamation of lands, the improvement of agriculture, and the building of bridges were mainly due to the skill and industry of the old churchmen. When their ecclesiastical organization was destroyed, the country speedily relapsed into the state from which they had raised it; and Scotland continued to lie in ruins almost till our own day, when it has again been rescued from barrenness, more effectually even than before, by the combined influences of roads, education, and industry.

* Professor Innes's 'Sketches of Early Scottish History.' The principal ancient bridges in Scotland were those over the Tay at Perth (erected in the thirteenth century); over the Esk at Brechin and Marykirk; over the Dee at Kincardine, O'Neil, and Aberdeen; over the Don, near the same city; over the Spey at Orkhill; over the Clyde at Glasgow; over the Forth at Stirling; and over the Tyne at Haddington.

CHAPTER V.

ROADS AND TRAVELLING IN ENGLAND TOWARDS THE END OF LAST CENTURY.

THE progress made in the improvement of the roads throughout England was exceedingly slow. Though some of the main throughfares were mended so as to admit of stage-coach travelling at the rate of from four to six miles an hour, the less frequented roads continued to be all but impassable. Travelling was still difficult, tedious, and dangerous. Only those who could not well avoid it ever thought of undertaking a journey, and travelling for pleasure was out of the question. A writer in the 'Gentleman's Magazine' in 1752 says that a Londoner at that time would no more think of travelling into the west of England for pleasure than of going to Nubia.

But signs of progress were not awanting. In 1749 Birmingham started a stage-coach, which made the journey to London in three days.* In 1754 some enterprising Manchester men advertised a "flying coach" for the conveyance of passengers between that town and the metro-

* Lady Luxborough, in a letter to Shenstone the poet, in 1749, says,—"A Birmingham coach is newly established to our great emolument. Would it not be a good scheme (this dirty weather, when riding is no more a pleasure) for you to come some Monday in the said stage-coach from Birmingham to breakfast at Barrells, (for they always breakfast at Henley); and on the Saturday following it would convey you back to Birmingham, unless you would stay longer, which would be better still, and equally easy; for the stage goes every week the same road. It breakfasts at Henley, and lies at Chipping Horton; goes early next day to Oxford, stays there all day and night, and gets on *the third day* to London; which from Birmingham at this season *is pretty well*, considering how long they are at Oxford; and it is much more agreeable as to the country than the Warwick way was."

polis; and, lest they should be classed with projectors of the Munchausen kind, they heralded their enterprise with this statement: "However incredible it may appear, this coach will actually (barring accidents) arrive in London in four days and a half after leaving Manchester!"

Fast coaches were also established on several of the northern roads, though not with very extraordinary results as to speed. When John Scott, afterwards Lord Chancellor Eldon, travelled from Newcastle to Oxford in 1766, he mentions that he journeyed in what was denominated "a fly," because of its rapid travelling; yet he was three or four days and nights on the road. There was no such velocity, however, as to endanger overturning or other mischief. On the panels of the coach were painted the appropriate motto of *Sat cito si sat bene*—quick enough if well enough—a motto which the future Lord Chancellor made his own.*

The journey by coach between London and Edinburgh still occupied six days or more, according to the state of the weather. Between Bath or Birmingham and London occupied between two and three days as late as 1763. The road across Hounslow Heath was so bad, that it was stated before a Parliamentary Committee that it was frequently known to be two feet deep in mud. The rate of travelling was about six and a half miles an hour; but the work was so heavy that it "tore the horses' hearts out," as the common saying went, so that they only lasted two or three years.

* We may incidentally mention three other journeys south by future Lords Chancellors. Mansfield rode up from Scotland to London when a boy, taking two months to make the journey on his pony. Wedderburn's journey by coach from Edinburgh to London, in 1757, occupied him six days. "When I first reached London," said the late Lord Campbell, "I performed the same journey in three nights and two days, Mr. Palmer's mail-coaches being then established; but this swift travelling was considered dangerous as well as wonderful, and I was gravely advised to stay a day at York, as several passengers who had gone through without stopping had died of apoplexy from the rapidity of the motion!"

When the Bath road became improved, Burke was enabled, in the summer of 1774, to travel from London to Bristol, to meet the electors there, in little more than four and twenty hours; but his biographer takes care to relate that he "travelled with incredible speed." Glasgow was still ten days' distance from the metropolis, and the arrival of the mail there was so important an event that a gun was fired to announce its coming in. Sheffield set up a "flying machine on steel springs" to London in 1760: it "slept" the first night at the Black Man's Head Inn, Nottingham; the second at the Angel, Northampton; and arrived at the Swan with Two Necks, Lad-lane, on the evening of the third day. The fare was *1l.* *17s.*, and 14 lbs. of luggage was allowed. But the principal part of the expense of travelling was for living and lodging on the road, not to mention the fees to guards and drivers.

Though the Dover road was still one of the best in the kingdom, the Dover flying-machine, carrying only four passengers, took a long summer's day to perform the journey. It set out from Dover at four o'clock in the morning, breakfasted at the Red Lion, Canterbury, and the passengers ate their way up to town at various inns on the road, arriving in London in time for supper. Smollett complained of the innkeepers along that route as the greatest set of extortioners in England. The deliberate style in which journeys were performed may be inferred from the circumstance that on one occasion, when a quarrel took place between the guard and a passenger, the coach stopped to see them fight it out on the road.

Foreigners who visited England were peculiarly observant of the defective modes of conveyance then in use. Thus, one Don Manoel Gonzales, a Portuguese merchant, who travelled through Great Britain in 1740, speaking of Yarmouth, says, "They have a comical way of carrying people all over the town and from the seaside, for six pence. They call it their coach, but it is only a wheel-barrow, drawn by one horse, without any covering." Another foreigner, Herr Alberti, a Hanoverian professor of theology, when on a visit to Oxford in 1750, desiring to proceed to Cambridge, found there was no means of doing so without returning to London and there taking coach for Cambridge. There was not even the convenience of a carrier's waggon between the two universities. But the most amusing account of an actual journey by stage-coach that we know of, is that given by a Prussian clergyman, Charles H. Moritz, who thus describes his adventures on the road between Leicester and London in 1782:—

"Being obliged," he says, "to bestir myself to get back to London, as the time drew near when the Hamburgh captain with whom I intended to return had fixed his departure, I determined to take a place as far as Northampton on the outside. But this ride from Leicester to Northampton I shall remember as long as I live.

"The coach drove from the yard through a part of the house. The inside passengers got in from the yard, but we on the outside were obliged to clamber up in the street, because we should have had no room for our heads to pass under the gateway. My companions on the top of the coach were a farmer, a young man very decently dressed, and a black-a-moor. The getting up alone was at the risk of one's life, and when I was up I was obliged to sit just at the corner of the coach, with nothing to hold by but a sort of little handle fastened on the side. I sat nearest the wheel, and the moment that we set off I fancied that I saw certain death before me. All I could do was to take still tighter hold of the handle, and to be strictly careful to preserve my balance. The machine rolled along with prodigious rapidity over the stones through the town, and every moment we seemed to fly into the air, so much so that it appeared to me a complete miracle that we stuck to the coach at all. But we were completely on the wing as often as we passed through a village or went down a hill.

"This continual fear of death at last became insupportable to me, and, therefore, no sooner were we crawling up a rather steep hill, and consequently proceeding slower than usual, then I carefully crept from the top of the coach, and was lucky enough to get myself snugly ensconced in the basket behind.

"'O, Sir, you will be shaken to death!' said the black-a-moor; but I heeded him not, trusting that he was exaggerating the unpleasantness of my new situation. And truly, as long as we went on slowly up the hill it was easy and pleasant enough; and I was just on the point of falling asleep among the surrounding trunks and packages, having had no rest the night before, when on a sudden the coach proceeded at a rapid rate down the hill. Then all the boxes, iron-nailed and copper-fastened, began, as it were, to dance around me; everything in the basket appeared to be alive, and every moment I received such violent blows that I thought my last hour had come. The black-a-moor had been right, I now saw clearly; but repentance was useless, and I was obliged to suffer horrible torture for nearly an hour, which seemed to me an eternity. At last we came to another hill, when, quite shaken to pieces, bleeding, and sore, I ruefully crept back to the top of the coach to my former seat. 'Ah, did I not tell you that you would be shaken to death?' inquired the black man, when I was creeping along on my stomach. But I gave him no reply. Indeed, I was ashamed; and I now write this as a warning to all strangers who are inclined to ride in English stage-coaches, and take an outside seat, or, worse still, horror of horrors, a seat in the basket.

"From Harborough to Northampton I had a most dreadful journey. It rained incessantly, and as before we had been covered with dust, so now we were soaked with rain. My neighbour, the young man who sat next me in the middle, every now and then fell asleep; and when in this state he perpetually bolted and rolled against me, with the whole weight of his body, more than once nearly pushing me from my seat, to which I clung with the last strength of despair. My forces were nearly giving way, when at last, happily, we reached Northampton, on the evening of the 14th July, 1782, an ever-memorable day to me.

"On the next morning, I took an *inside* place for London. We started early in the morning. The journey from Northampton to the metropolis, however, I can scarcely call a ride, for it was a perpetual motion, or endless jolt from one place to another, in a close wooden box, over what appeared to be a heap of unhewn stones and trunks of trees scattered by a hurricane. To make my happiness complete, I had three travelling companions, all farmers, who slept so soundly that even the hearty knocks with which they hammered their heads against each other and against mine did not awake them. Their faces, bloated and discoloured by ale and brandy and the knocks aforesaid, looked, as they lay before me, like so many lumps of dead flesh.

"I looked, and certainly felt, like a crazy fool when we arrived at London in the afternoon." *

THE BASKET COACH, 1780. [By Louis Huard, after Rowlandson.]

Arthur Young, in his books, inveighs strongly against the execrable state of the roads in all parts of England towards the end of last century. In Essex he found the ruts "of an incredible depth," and he almost swore at one near Tilbury. "Of all the cursed roads," he says, "that ever disgraced this kingdom in the very ages of barbarism, none ever equalled that from Billericay to the King's Head at Tilbury. It is for near twelve miles so narrow that a mouse cannot pass by any carriage. I saw a fellow creep under his waggon to assist me to lift, if possible, my chaise over a hedge. To add to all the infamous circumstances

* C. H. Moritz: 'Reise eines Deutschen in England im Jahre 1782.' Berlin, 1783.

which concur to plague a traveller, I must not forget the eternally meeting with chalk waggons, themselves frequently stuck fast, till a collection of them are in the same situation, and twenty or thirty horses may be tacked to each to draw them out one by one!" * Yet will it be believed, the proposal to form a turnpike-road from Chelmsford to Tilbury was resisted "by the Bruins of the country, whose horses were worried to death with bringing chalk through those vile roads!"

Arthur Young did not find the turnpike any better between Bury and Sudbury, in Suffolk: "I was forced to move as slow in it," he says, "as in any unmended lane in Wales. For, ponds of liquid dirt, and a scattering of loose flints just sufficient to lame every horse that moves near them, with the addition of cutting vile grips across the road under the pretence of letting the water off, but without effect, altogether render at least twelve out of these sixteen miles as infamous a turnpike as ever was beheld." Between Tetsworth and Oxford he found the so-called turnpike abounding in loose stones as large as one's head, full of holes, deep ruts, and withal so narrow that with great difficulty he got his chaise out of the way of the Witney waggons. "Barbarous" and "execrable" are the words which he constantly employs in speaking of the roads; parish and turnpike, all seemed to be alike bad. From Gloucester to Newnham, a distance of twelve miles, he found a "cursed road," "infamously stony," with "ruts all the way." From Newnham to Chepstow he noted another bad feature in the roads, and that was the perpetual hills; "for," he says, "you will form a clear idea of them if you suppose the country to represent the roofs of houses joined, and the road to run across them." It was at one time even matter of grave dispute whether it would not cost as little money to make that between Leominster and

* Arthur Young's 'Six Weeks' Tour through the Southern Counties of England and Wales.' 2nd ed., 1769, pp. 88-9.

Kington navigable as to make it hard. Passing still further west, the unfortunate traveller, who seems scarcely able to find words to express his sufferings, continues:—

"But, my dear Sir, what am I to say of the roads in this country! the turnpikes! as they have the assurance to call them and the hardiness to make one pay for? From Chepstow to the half-way house between Newport and Cardiff they continue mere rocky lanes, full of hugeous stones as big as one's horse, and abominable holes. The first six miles from Newport they were so detestable, and without either direction-posts or milestones, that I could not well persuade myself I was on the turnpike, but had mistook the road, and therefore asked every one I met, who answered me, to my astonishment, '*Ya-as!*' Whatever business carries you into this country, avoid it, at least till they have good roads: if they were good, travelling would be very pleasant." *

At a subsequent period Arthur Young visited the northern counties; but his account of the roads in that quarter is not more satisfactory. Between Richmond and Darlington he found them like to "dislocate his bones," being broken in many places into deep holes, and almost impassable; "yet," says he, "the people *will* drink tea!"—a decoction against the use of which the traveller is found constantly declaiming. The roads in Lancashire made him almost frantic, and he gasped for words to express his rage. Of the road between Proud Preston and Wigan he says: "I know not in the whole range of language terms sufficiently expressive to describe this infernal road. Let me most seriously caution all travellers who may accidentally propose to travel this terrible country, to avoid it as they would the devil; for a thousand to one they break their necks or their limbs by overthrows or breakings-down.

* 'Six Weeks' Tour in the Southern Counties of England and Wales,' pp. 153-5. The roads all over South Wales were equally bad down to the beginning of the present century. At Halfway, near Trecastle, in Breconshire, South Wales, a small obelisk is still to be seen, which was erected to commemorate the turn over and destruction of the mail coach over a steep of 130 feet; the driver and passengers escaping unhurt.

They will here meet with ruts, which I actually measured, *four feet deep*, and floating with mud only from a wet summer. What, therefore, must it be after a winter? The only mending it receives is tumbling in some loose stones, which serve no other purpose than jolting a carriage in the most intolerable manner. These are not merely opinions, but facts; for I actually passed *three carts broken down* in those eighteen miles of execrable memory." *

It would even appear that the bad state of the roads in the Midland counties, about the same time, had nearly caused the death of the heir to the throne. On the 2nd of September, 1789, the Prince of Wales left Wentworth Hall, where he had been on a visit to Earl Fitzwilliam, and took the road for London in his carriage. When about two miles from Newark the Prince's coach was overturned by a cart in a narrow part of the road; it rolled down a slope, turning over three times, and landed at the bottom, shivered to pieces. Fortunately the Prince escaped with only a few bruises and a sprain; but the incident had no effect in stirring up the local authorities to make any improvement in the road, which remained in the same wretched state until a comparatively recent period.

When Palmer's new mail-coaches were introduced, an attempt was made to diminish the jolting of the passengers by having the carriages hung upon new patent springs, but with very indifferent results. Mathew Boulton, the engineer, thus described their effect upon himself in a journey he made in one of them from London into Devonshire, in 1787:—

"I had the most disagreeable journey I ever experienced the night after I left you, owing to the new improved patent coach, a vehicle loaded with iron trappings and the greatest complication of unmechanical contrivances jumbled together, that I have ever witnessed. The coach swings sideways, with a sickly sway without

* 'A Six Months' Tour through the North of England,' vol. iv. p. 431.

any vertical spring; the point of suspense bearing upon an arch called a spring, though it is nothing of the sort. The severity of the jolting occasioned me such disorder, that I was obliged to stop at Axminster and go to bed very ill. However, I was able next day to proceed in a post-chaise. The landlady in the London Inn, at Exeter, assured me that the passengers who arrived every night were in general so ill that they were obliged to go supperless to bed; and, unless they go back to the old-fashioned coach, hung a little lower, the mail-coaches will lose all their custom." *

We may briefly refer to the several stages of improvement—if improvement it could be called—in the most frequented highways of the kingdom, and to the action of the legislature with reference to the extension of turnpikes. The trade and industry of the country had been steadily improving; but the greatest obstacle to their further progress was always felt to be the disgraceful state of the roads. As long ago as the year 1663 an Act was passed † authorising the first toll-gates or turnpikes to be erected, at which collectors were stationed to levy small sums from those using the road, for the purpose of defraying the needful expenses of their maintenance. This Act, however, only applied to a portion of the Great North Road between London and York, and it authorised the new toll-bars to be erected at Wade's Mill in Hertfordshire, at Caxton in Cambridgeshire, and at Stilton in Huntingdonshire.‡ The Act was not followed by any others for a

* Letter to Wyatt, October 5th, 1787, MS.

† Act 15 Car. II., c. 1.

‡ The preamble of the Act recites that "The ancient highway and post-road leading from London to York, and so into Scotland, and likewise from London into Lincolnshire, lieth for many miles in the counties of Hertford, Cambridge, and Huntingdon, in many of which places the road, by reason of the great and many loads which are weekly drawn in waggons through the said places, as well as by reason of the great trade of barley and malt that cometh to Ware, and so is conveyed by water to the city of London, as well as other carriages, both from the north parts as also from the city of Norwich, St. Edmondsbury, and the town of Cambridge, to London, is very ruinous, and become almost impassable, insomuch that it is become very dangerous to all his Majesty's liege people that pass that way," &c.

quarter of a century, and even after that lapse of time such Acts as were passed of a similar character were very few and far between.

For nearly a century more, travellers from Edinburgh to London met with no turnpikes until within about 110 miles of the metropolis. North of that point there was only a narrow causeway fit for pack-horses, flanked with clay sloughs on either side. It is, however, stated that the Duke of Cumberland and the Earl of Albemarle, when on their way to Scotland in pursuit of the rebels in 1746, did contrive to reach Durham in a coach and six; but there the roads were found so wretched, that they were under the necessity of taking to horse, and Mr. George Bowes, the county member, made His Royal Highness a present of his nag to enable him to proceed on his journey. The roads west of Newcastle were so bad, that in the previous year the royal forces under General Wade, which left Newcastle for Carlisle to intercept the Pretender and his army, halted the first night at Ovingham, and the second at Hexham, being able to travel only twenty miles in two days.*

The rebellion of 1745 gave a great impulse to the construction of roads for military as well as civil purposes. The nimble Highlanders, without baggage or waggons, had been able to cross the border and penetrate almost to the centre of England before any definite knowledge of their proceedings had reached the rest of the kingdom. In the metropolis itself little information could be obtained of the movements of the rebel army for several days after they had left Edinburgh. Light of foot, they outstripped the cavalry and artillery of the royal army, which were delayed at all points by impassable

* Down to the year 1756, Newcastle and Carlisle were only connected by a bridle way. In that year, Marshal Wade employed his army to construct a road by way of Harlaw and Cholterford, following for thirty miles the line of the old Roman Wall, the materials of which he used to construct his "agger" and culverts. This was long after known as "the military road."

roads. No sooner, however, was the rebellion put down, than Government directed its attention to the best means of securing the permanent subordination of the Highlands, and with this object the construction of good highways was declared to be indispensable. The expediency of opening up the communication between the capital and the principal towns of Scotland was also generally admitted; and from that time, though slowly, the construction of the main high routes between north and south made steady progress.

The extension of the turnpike system, however, encountered violent opposition from the people, being regarded as a grievous tax upon their freedom of movement from place to place. Armed bodies of men assembled to destroy the turnpikes; and they burnt down the toll-houses and blew up the posts with gunpowder. The resistance was the greatest in Yorkshire, along the line of the Great North Road towards Scotland, though riots also took place in Somersetshire and Gloucestershire, and even in the immediate neighbourhood of London. One fine May morning, at Selby, in Yorkshire, the public bellman summoned the inhabitants to assemble with their hatchets and axes that night at midnight, and cut down the turnpikes erected by Act of Parliament; nor were they slow to act upon his summons. Soldiers were then sent into the district to protect the toll-bars and the toll-takers; but this was a difficult matter, for the toll-gates were numerous, and wherever a "pike" was left unprotected at night, it was found destroyed in the morning. The Yeadon and Otley mobs, near Leeds, were especially violent. On the 18th of June, 1753, they made quite a raid upon the turnpikes, burning or destroying about a dozen in one week. A score of the rioters were apprehended, and while on their way to York Castle a rescue was attempted, when the soldiers were under the necessity of firing, and many persons were killed and wounded.

The prejudices entertained against the turnpikes were

so strong, that in some places the country people would not even use the improved roads after they were made.* For instance, the driver of the Marlborough coach obstinately refused to use the New Bath road, but stuck to the old waggon-track, called "Ramsbury." He was an old man, he said: his grandfather and father had driven the aforesaid way before him, and he would continue in the old track till death.†

Petitions were also presented to Parliament against the extension of turnpikes; but the opposition represented by the petitioners was of a much less honest character than that of the misguided and prejudiced country folks, who burnt down the toll-houses. It was principally got up by the agriculturists in the neighbourhood of the metropolis, who, having secured the advantages which the turnpike-roads first constructed had conferred upon them, desired to retain a monopoly of the improved means of communication. They alleged that if turnpike-roads were extended into the remoter counties, the greater cheapness of labour there would enable the distant farmers to sell their grass and corn cheaper in the London market than themselves, and that thus they would be ruined.‡ This opposition, however, did not prevent the progress of turnpike and highway legislation; and we find that, from 1760 to 1774, no fewer than four hundred and fifty-two Acts were passed for making and repairing highways. Nevertheless the roads of the kingdom long continued in a very unsatisfactory state, chiefly arising from the extremely imperfect manner in which they were made.

Road-making as a profession was as yet unknown. Deviations were made in the old roads to make them more

* The Blandford waggoner said, "Roads had but one object—for waggon-driving. He required but four-foot width in a lane, and all the rest might go to the devil." He added, "The gentry ought to stay at home, and be d——d, and not run gossiping up and down the country."—Roberts's 'Social History of the Southern Counties.'

† 'Gentleman's Magazine' for December, 1752.

‡ Adam Smith's 'Wealth of Nations,' book i., chap. xi., part i.

easy and straight; but the deep ruts were merely filled up with any materials that lay nearest at hand, and stones taken from the quarry, instead of being broken and laid on carefully to a proper depth, were tumbled down and roughly spread, the country road-maker trusting to the operation of cart-wheels and waggons to crush them into a proper shape. Men of eminence as engineers—and there were very few such at the time—considered road-making beneath their consideration; and it was even thought singular that, in 1768, the distinguished Smeaton should have condescended to make a road across the valley of the Trent, between Markham and Newark.

The making of the new roads was thus left to such persons as might choose to take up the trade, special skill not being thought at all necessary on the part of a road-maker. It is only in this way that we can account for the remarkable fact, that the first extensive maker of roads who pursued it as a business, was not an engineer, nor even a mechanic, but a Blind Man, bred to no trade, and possessing no experience whatever in the arts of surveying or bridge-building, yet a man possessed of extraordinary natural gifts, and unquestionably most successful as a road-maker. We allude to John Metcalf, commonly known as "Blind Jack of Knaresborough," to whose biography, as the constructor of nearly two hundred miles of capital roads—as, indeed, the first great English road-maker—we propose to devote the next chapter.

CHAPTER VI.

JOHN METCALF, ROAD-MAKER.

JOHN METCALF was born at Knaresborough in 1717, the son of poor working people. When only six years old he was seized with virulent small-pox, which totally destroyed his sight. The blind boy, when sufficiently recovered to go abroad, first learnt to grope from door to door along the walls on either side of his parents' dwelling. In about six months he was able to feel his way to the end of the street and back without a guide, and in three years he could go on a message to any part of the town. He grew strong and healthy, and longed to join in the sports of boys of his age. He went bird-nesting with them, and climbed the trees while the boys below directed him to the nests, receiving his share of eggs and young birds. Thus he shortly became an expert climber, and could mount with ease any tree that he was able to grasp. He rambled into the lanes and fields alone, and soon knew every foot of the ground for miles round Knaresborough. He next learnt to ride, delighting above all things in a gallop. He contrived to keep a dog and coursed hares: indeed, the boy was the marvel of the neighbourhood. His unrestrainable activity, his acuteness of sense, his shrewdness, and his cleverness, astonished everybody.

The boy's confidence in himself was such, that though blind, he was ready to undertake almost any adventure. Among his other arts he learned to swim in the Nidd, and became so expert that on one occasion he saved the lives of three of his companions. Once, when two men were drowned in a deep part of the river, Metcalf was sent for to dive for them, which he did, and brought up one of the

METCALF'S BIRTHPLACE, KNARESBOROUGH.

By E. M. WIMPRIS, after SUTCLIFF.

Page 74.

bodies at the fourth diving: the other had been carried down the stream. He thus also saved a manufacturer's yarn, a large quantity of which had been carried by a sudden flood into a deep hole under the High Bridge. At home, in the evenings, he learnt to play the fiddle, and became so skilled on the instrument, that he was shortly able to earn money by playing dance music at country parties. At Christmas time he played waits, and during the Harrogate season he played to the assemblies at the Queen's Head and the Green Dragon.

On one occasion, towards dusk, he acted as guide to a belated gentleman along the difficult road from York to Harrogate. The road was then full of windings and turnings, and in many places it was no better than a track across unenclosed moors. Metcalf brought the gentleman safe to his inn, "The Granby," late at night, and was invited to join in a tankard of negus. On Metcalf leaving the room, the gentleman observed to the landlord—"I think, landlord, my guide must have drunk a great deal of spirits since we came here." "Why so, Sir?" "Well, I judge so, from the appearance of *his eyes*." "Eyes! bless you, Sir," rejoined the landlord, "don't you know that he is *blind?*" "Blind! What do you mean by that?" "I mean, Sir, that he cannot see—he is as blind as a stone." "Well, landlord," said the gentleman, "this is really too much: call him in." Enter Metcalf. "My friend, are you *really* blind?" "Yes, Sir," said he, "I lost my sight when six years old." "Had I known that, I would not have ventured with you on that road from York for a hundred pounds." "And I, Sir," said Metcalf, "would not have lost my way for a thousand."

Metcalf having thriven and saved money, bought and rode a horse of his own. He had a great affection for the animal, and when he called, it would immediately answer him by neighing. The most surprising thing is that he was a good huntsman; and to follow the hounds was one of his greatest pleasures. He was as bold a rider as ever took the field. He trusted much, no doubt, to the sagacity of his horse; but he himself was apparently regardless of danger. The hunting adventures which are related of him, considering his blindness, seem altogether marvellous. He would also run his horse for the petty prizes or plates given at the "feasts" in the neighbourhood, and he attended the races at York and other places, where he made bets with considerable skill, keeping well in his memory the winning and losing horses. After the races, he would return to Knaresborough late at night, guiding others who but for him could never have made out the way.

On one occasion he rode his horse in a match in Knaresborough Forest. The ground was marked out by posts, including a circle of a mile, and the race was three times round. Great odds were laid against the blind man, because of his supposed inability to keep the course. But his ingenuity was never at fault. He procured a number of dinner-bells from the Harrogate inns and set men to ring them at the several posts. Their sound was enough to direct him during the race, and the blind man came in the winner! After the race was over, a gentleman who owned a notorious runaway horse came up and offered to lay a bet with Metcalf that he could not gallop the horse fifty yards and stop it within two hundred. Metcalf accepted the bet, with the condition that he might choose his ground. This was agreed to, but there was to be neither hedge nor wall in the distance. Metcalf forthwith proceeded to the neighbourhood of the large bog near the Harrogate Old Spa, and having placed a person on the line in which he proposed to ride, who was to sing a song to guide him by its sound, he mounted and rode straight into the bog, where he had the horse effectually stopped within the stipulated two hundred yards, stuck up to his saddle-girths in the mire. Metcalf scrambled out and claimed his wager; but it was with the greatest difficulty that the horse could be extricated.

The blind man also played at bowls very successfully, receiving the odds of a bowl extra for the deficiency of each eye. He had thus three bowls for the other's one; and he took care to place one friend at the jack and another midway, who, keeping up a constant discourse with him, enabled him readily to judge of the distance. In athletic sports, such as wrestling and boxing, he was also a great adept; and being now a full-grown man, of great strength and robustness, about six feet two in height, few durst try upon him the practical jokes which cowardly persons are sometimes disposed to play upon the blind.

Notwithstanding his mischievous tricks and youthful wildness, there must have been something exceedingly winning about the man, possessed, as he was, of a strong, manly, and affectionate nature; and we are not, therefore, surprised to learn that the landlord's daughter of "The Granby" fairly fell in love with Blind Jack and married him, much to the disgust of her relatives. When asked how it was that she could marry such a man, her womanlike reply was, "Because I could not be happy without him: his actions are so singular, and his spirit so manly and enterprising, that I could not help loving him." But, after all, Dolly was not so far wrong in the choice as her parents thought her. As the result proved, Metcalf had in him elements of success in life, which, even according to the world's estimate, made him eventually a very "good match," and the woman's clear sight in this case stood her in good stead.

But before this marriage was consummated, Metcalf had wandered far and "seen" a good deal of the world, as he termed it. He travelled on horseback to Whitby, and from thence he sailed for London, taking with him his fiddle, by the aid of which he continued to earn enough to maintain himself for several weeks in the metropolis. Returning to Whitby, he sailed from thence to Newcastle to "see" some friends there, whom he had known at Harrogate while visiting that watering-place. He was welcomed by many families and spent an agreeable month, afterwards visiting Sunderland, still supporting himself by his violin playing. Then he returned to Whitby for his horse, and rode homeward alone to Knaresborough by Pickering, Malton, and York, over very bad roads, the greater part of which he had never travelled before, yet without once missing his way. When he arrived at York, it was the dead of night, and he found the city gates at Middlethorp shut. They were of strong planks, with iron spikes fixed on the top; but throwing his horse's bridle-rein over one of the spikes, he climbed up, and by the help of a corner of the wall that joined the gates, he got safely over: then opening them from the inside, he led his horse through.

After another season at Harrogate, he made a second visit to London, in the company of a North countryman who played the small pipes. He was kindly entertained by Colonel Liddell, of Ravensworth Castle, who gave him a general invitation to his house. During this visit, which was in 1730-1, Metcalf ranged freely over the metropolis, visiting Maidenhead and Reading, and returning by Windsor and Hampton Court. The Harrogate season being at hand, he prepared to proceed thither,—Colonel Liddell, who was also about setting out for Harrogate, offering him a seat behind his coach. Metcalf thanked him, but declined the offer, observing that he could, with great ease, walk as far in a day as he, the Colonel, was likely to travel in his carriage; besides, he preferred the walking. That a blind man should undertake to walk a distance of two hundred miles over an unknown road, in the same time that it took a gentleman to perform the same distance in his coach, dragged by post-horses, seems almost incredible; yet Metcalf actually arrived at Harrogate before the Colonel, and that without hurrying by the way. The circumstance is easily accounted for by the deplorable state of the roads, which made travelling by foot on the whole considerably more expeditious than travelling by coach. The story is even extant of a man with a wooden leg being once offered a lift upon a stage-coach; but he declined, with "Thank'ee, I can't wait; I'm in a hurry." And he stumped on, ahead of the coach.

The account of Metcalf's journey on foot from London to Harrogate is not without a special bearing on our subject, as illustrative of the state of the roads at the time. He started on a Monday morning, about an hour before the Colonel in his carriage, with his suite, which consisted of sixteen servants on horseback. It was arranged that they should sleep that night at Welwyn, in Hertfordshire. Metcalf made his way to Barnet; but a little north of that town, where the road branches off to St. Albans, he took the wrong way, and thus made a considerable détour. Nevertheless he arrived at Welwyn first, to the surprise of the Colonel. Next morning he set off as before, and reached Biggleswade; but there he found the river swollen and no bridge provided to enable travellers to cross to the further side. He made a considerable circuit, in the hope of finding some method of crossing the stream, and was so fortunate as to fall in with a fellow wayfarer, who led the way across some planks, Metcalf following the sound of his feet. Arrived at the other side, Metcalf, taking some pence from his pocket, said, "Here, my good fellow, take that and get a pint of beer." The stranger declined, saying he was welcome to his services. Metcalf, however, pressed upon his guide the small reward, when the other asked, "Pray, can you see very well?" "Not remarkably well," said Metcalf. "My friend," said the stranger, "I do not mean to tithe you: I am the rector of this parish; so God bless you, and I wish you a good journey." Metcalf set forward again with the blessing, and reached his journey's end safely, again before the Colonel. On the Saturday after their setting out from London, the travellers reached Wetherby, where Colonel Liddell desired to rest until the Monday; but Metcalf proceeded on to Harrogate, thus completing the journey in six days,—the Colonel arriving two days later.

He now renewed his musical performances at Harrogate, and was also in considerable request at the Ripon assemblies, which were attended by most of the families of distinction in that neighbourhood. When the season at Harrogate was over, he retired to Knaresborough with his young wife, and having purchased an old house, he had it pulled down and another built on its site,—he himself getting the requisite stones for the masonry out of the bed of the adjoining river. The uncertainty of the income derived from musical performances led him to think of following some more settled pursuit, now that he had a wife to maintain as well as himself. He accordingly set up a four-wheeled and a one-horse chaise for the public accommodation,—Harrogate up to that time being without any vehicle for hire. The innkeepers of the town having followed his example, and abstracted most of his business, Metcalf next took to fish-dealing. He bought fish at the coast, which he conveyed on horseback to Leeds and other towns for sale. He continued indefatigable at this trade for some time, being on the road often for nights together; but he was at length forced to abandon it in consequence of the inadequacy of the returns. He was therefore under the necessity of again taking up his violin; and he was employed as a musician in the Long Room at Harrogate, at the time of the outbreak of the Rebellion of 1745.

The news of the rout of the Royal army at Prestonpans, and the intended march of the Highlanders southwards, put a stop to business as well as pleasure, and caused a general consternation throughout the northern counties. The great bulk of the people were, however, comparatively indifferent to the measures of defence which were adopted; and but for the energy displayed by the country gentlemen in raising forces in support of the established government, the Stuarts might again have been seated on the throne of Britain. Among the county gentlemen of York who distinguished themselves on the occasion was William Thornton, Esq., of Thornville Royal. The county having voted ninety thousand pounds for raising, clothing, and maintaining a body of four thousand men, Mr. Thornton proposed, at a public meeting held at York, that they should be embodied with the regulars and march with the King's forces to meet the Pretender in the field. This proposal was, however, overruled, the majority of the meeting resolving that the men should be retained at home for purposes merely of local defence. On this decision being come to, Mr. Thornton determined to raise a company of volunteers at his own expense, and to join the Royal army with such force as he could muster. He then went abroad among his tenantry and servants, and endeavoured to induce them to follow him, but without success.

Still determined on raising his company, Mr. Thornton next cast about him for other means; and who should he think of in his emergency but Blind Jack! Metcalf had often played to his family at Christmas time, and the Squire knew him to be one of the most popular men in the neighbourhood. He accordingly proceeded to Knaresborough to confer with Metcalf on the subject. It was then about the beginning of October, only a fortnight after the battle of Prestonpans. Sending for Jack to his inn, Mr. Thornton told him of the state of affairs—that the French were coming to join the rebels—and that if the country were allowed to fall into their hands, no man's wife, daughter, nor sister would be safe. Jack's loyalty was at once kindled. If no one else would join the Squire, he would! Thus enlisted—perhaps carried away by his love of adventure not less than by his feeling of patriotism—Metcalf proceeded to enlist others, and in two days a hundred and forty men were obtained, from whom Mr. Thornton drafted sixty-four, the intended number of his company. The men were immediately drilled and brought into a state of as much efficiency as was practicable in the time; and when they marched off to join General Wade's army at Boroughbridge, the Captain said to them on setting out, "My lads! you are going to form part of a ring-fence to the finest estate in the world!" Blind Jack played a march at the head of the company, dressed in blue and buff, and in a gold-laced hat. The Captain said he would willingly give a hundred guineas for only one eye to put in Jack's head: he was such a useful, spirited, handy fellow.

On arriving at Newcastle, Captain Thornton's company was united to Pulteney's regiment, one of the weakest. The army lay for a week in tents on the Moor. Winter had set in, and the snow lay thick on the ground; but intelligence arriving that Prince Charles, with his Highlanders, was proceeding southwards by way of Carlisle, General Wade gave orders for the immediate advance of the army on Hexham, in the hope of intercepting them by that route. They set out on their march amidst hail and snow; and in addition to the obstruction caused by the weather, they had to overcome the difficulties occasioned by the badness of the roads. The men were often three or four hours in marching a mile, the pioneers having to fill up ditches and clear away many obstructions in making a practicable passage for the artillery and baggage. The army was only able to reach Ovingham, a distance of little more than ten miles, after fifteen hours' marching. The night was bitter cold; the ground was frozen so hard that but few of the tent-pins could be driven; and the men lay down upon the earth amongst their straw. Metcalf, to keep up the spirits of his company—for sleep was next to impossible—took out his fiddle and played lively tunes whilst the men danced round the straw, which they set on fire.

Next day the army marched for Hexham; but the rebels having already passed southward, General Wade retraced his steps to Newcastle to gain the high road leading to Yorkshire, whither he marched in all haste; and for a time his army lay before Leeds on fields now covered with streets, some of which still bear the names of Wade-lane, Camp-road, and Camp-field, in consequence of the event.

On the retreat of Prince Charles from Derby, General Wade again proceeded to Newcastle, while the Duke of Cumberland hung upon the rear of the rebels along their line of retreat by Penrith and Carlisle. Wade's army proceeded by forced marches into Scotland, and at length came up with the Highlanders at Falkirk. Metcalf continued with Captain Thornton and his company throughout all these marchings and countermarchings, determined to be of service to his master if he could, and at all events to see the end of the campaign. At the battle of Falkirk he played his company to the field; but it was a grossly-mismanaged battle on the part of the Royalist General, and the result was a total defeat. Twenty of Thornton's men were made prisoners, with the lieutenant and ensign. The Captain himself only escaped by taking refuge in a poor woman's house in the town of Falkirk, where he lay hidden for many days; Metcalf returning to Edinburgh with the rest of the defeated army.

Some of the Dragoon officers, hearing of Jack's escape, sent for him to head-quarters at Holyrood, to question him about his Captain. One of them took occasion to speak ironically of Thornton's men, and asked Metcalf how *he* had contrived to escape. "Oh!" said Jack, "I found it easy to follow the sound of the Dragoons' horses—they made such a clatter over the stones when flying from the Highlandmen." Another asked him how he, a blind man, durst venture upon such a service; to which Metcalf replied, that had he possessed a pair of good eyes, perhaps he would not have come there to risk the loss of them by gunpowder. No more questions were asked, and Jack withdrew; but he was not satisfied about the disappearance of Captain Thornton, and determined on going back to Falkirk, within the enemy's lines, to get news of him, and perhaps to rescue him, if that were still possible.

The rest of the company were very much disheartened at the loss of their officers and so many of their comrades, and wished Metcalf to furnish them with the means of returning home. But he would not hear of such a thing, and strongly encouraged them to remain until, at all events, he had got news of the Captain. He then set out for Prince Charles's camp. On reaching the outposts of the English army, he was urged by the officer in command to lay aside his project, which would certainly cost him his life. But Metcalf was not to be dissuaded, and he was permitted to proceed, which he did in the company of one of the rebel spies, pretending that he wished to be engaged as a musician in the Prince's army. A woman whom they met returning to Edinburgh from the field of Falkirk, laden with plunder, gave Metcalf a token to her husband, who was Lord George Murray's cook, and this secured him an access to the Prince's quarters; but, notwithstanding a most diligent search, he could hear nothing of his master. Unfortunately for him, a person who had seen him at Harrogate, pointed him out as a suspicious character, and he was seized and put in confinement for three days, after which he was tried by court martial; but as nothing could be alleged against him, he was acquitted, and shortly after made his escape from the rebel camp. On reaching Edinburgh, very much to his delight he found Captain Thornton had arrived there before him.

On the 30th of January, 1746, the Duke of Cumberland reached Edinburgh, and put himself at the head of the Royal army, which proceeded northward in pursuit of the Highlanders. At Aberdeen, where the Duke gave a ball, Metcalf was found to be the only musician in camp who could play country dances, and he played to the company, standing on a chair, for eight hours,—the Duke several times, as he passed him, shouting out "Thornton, play up!" Next morning the Duke sent him a present of two guineas; but as the Captain would not allow him to receive such gifts while in his pay, Metcalf spent the money, with his permission, in giving a treat to the Duke's two body servants. The battle of Culloden, so disastrous to the poor Highlanders, shortly followed; after which Captain Thornton, Metcalf, and the Yorkshire Volunteer Company, proceeded homewards. Metcalf's young wife had been in great fears for the safety of her blind, fearless, and almost reckless partner; but she received him with open arms, and his spirit of adventure being now considerably allayed, he determined to settle quietly down to the steady pursuit of business.

During his stay in Aberdeen, Metcalf had made himself familiar with the articles of clothing manufactured at that place, and he came to the conclusion that a profitable trade might be carried on by buying them on the spot and selling them by retail to customers in Yorkshire. He accordingly proceeded to Aberdeen in the following spring, and bought a considerable stock of cotton and worsted stockings, which he found he could readily dispose of on his return home. His knowledge of horseflesh—in which he was, of course, mainly guided by his acute sense of feeling—also proved highly serviceable to him, and he bought considerable numbers of horses in Yorkshire for sale in Scotland, bringing back galloways in return. It is supposed that at the same time he carried on a profitable contraband trade in tea and such like articles.

After this, Metcalf began a new line of business, that of common carrier between York and Knaresborough, plying the first stage-waggon on that road. He made the journey twice a week in summer and once a week in winter. He also undertook the conveyance of army baggage, most other owners of carts at that time being afraid of soldiers, regarding them as a wild rough set, with whom it was dangerous to have any dealings. But the blind man knew them better, and while he drove a profitable trade in carrying their baggage from town to town, they never did him any harm. By these means, he very shortly succeeded in realising a considerable store of savings, besides being able to maintain his family in respectability and comfort.

Metcalf, however, had not yet entered upon the main business of his life. The reader will already have observed how strong of heart and resolute of purpose he was. During his adventurous career he had acquired a more than ordinary share of experience of the world. Stone blind as he was from his childhood, he had not been able to study books, but he had carefully studied men. He could read characters with wonderful quickness, rapidly taking stock, as he called it, of those with whom he came in contact. In his youth, as we have seen, he could follow the hounds on horse or on foot, and managed to be in at the death with the most expert riders. His travels about the country as a guide to those who could see, as a musician, soldier, chapman, fish-dealer, horse-dealer, and waggoner, had given him a perfectly familiar acquaintance with the northern roads. He could measure timber or hay in the stack, and rapidly reduce their contents to feet and inches after a mental process of his own. Withal he was endowed with an extraordinary activity and spirit of enterprise, which, had his sight been spared him, would probably have rendered him one of the most extraordinary men of his age. As it was, Metcalf now became one of the greatest of its road-makers and bridge-builders.

About the year 1765 an Act was passed empowering a turnpike-road to be constructed between Harrogate and Boroughbridge. The business of contractor had not yet come into existence, nor was the art of road-making much understood; and in a remote country place such as Knaresborough the surveyor had some difficulty in finding persons capable of executing the necessary work. The shrewd Metcalf discerned in the proposed enterprise the first of a series of public roads of a similar kind throughout the northern counties, for none knew better than he did how great was the need of them. He determined, therefore, to enter upon this new line of business, and offered to Mr. Ostler, the master surveyor, to construct three miles of the proposed road between Minskip and Fearnsby. Ostler knew the man well, and having the greatest confidence in his abilities, he let him the contract. Metcalf sold his

JOHN METCALF, THE BLIND ROAD-MAKER.

Page 86.

stage-waggons and his interest in the carrying business between York and Knaresborough, and at once proceeded with his new undertaking. The materials for metaling the road were to be obtained from one gravel-pit for the whole length, and he made his arrangements on a large scale accordingly, hauling out the ballast with unusual expedition and economy, at the same time proceeding with the formation of the road at all points; by which means he was enabled the first to complete his contract, to the entire satisfaction of the surveyor and trustees.

This was only the first of a vast number of similar projects on which Metcalf was afterwards engaged, extending over a period of more than thirty years. By the time that he had finished the road, the building of a bridge at Boroughbridge was advertised, and Metcalf sent in his tender with many others. At the same time he frankly stated that, though he wished to undertake the work, he had not before executed anything of the kind. His tender being on the whole the most favourable, the trustees sent for Metcalf, and on his appearing before them, they asked him what he knew of a bridge. He replied that he could readily describe his plan of the one they proposed to build, if they would be good enough to write down his figures. "The span of the arch, 18 feet," said he, "being a semicircle, makes 27: the arch-stones must be a foot deep, which, if multiplied by 27, will be 486; and the basis will be 72 feet more. This for the arch; but it will require good backing, for which purpose there are proper stones in the old Roman wall at Aldborough, which may be used for the purpose, if you please to give directions to that effect." It is doubtful whether the trustees were able to follow his rapid calculations; but they were so much struck by his readiness and apparently complete knowledge of the work he proposed to execute, that they gave him the contract to build the bridge; and he completed it within the stipulated time in a satisfactory and workmanlike manner.

He next agreed to make the mile and a half of turnpike-road between his native town of Knaresborough and Harrogate—ground with which he was more than ordinarily familiar. Walking one day over a portion of the ground on which the road was to be made, while still covered with grass, he told the workmen that he thought it differed from the ground adjoining it, and he directed them to try for stone or gravel underneath; and, strange to say, not many feet down, the men came upon the stones of an old Roman causeway, from which he obtained much valuable material for the making of his new road. At another part of the contract there was a bog to be crossed, and the surveyor thought it impossible to make a road over it. Metcalf assured him that he could readily accomplish it; on which the other offered, if he succeeded, to pay him for the straight road the price which he would have to pay if the road were constructed round the bog. Metcalf set to work accordingly, and had a large quantity of furze and ling laid upon the bog, over which he spread layers of gravel. The plan answered effectually, and when the materials had become consolidated, it proved one of the best parts of the road.

It would be tedious to describe in detail the construction of the various roads and bridges which Metcalf subsequently executed, but a brief summary of the more important will suffice. In Yorkshire, he made the roads between Harrogate and Harewood Bridge; between Chapeltown and Leeds; between Broughton and Addingham; between Mill Bridge and Halifax; between Wakefield and Dewsbury; between Wakefield and Doncaster; between Wakefield, Huddersfield, and Saddleworth (the Manchester road); between Standish and Thurston Clough; between Huddersfield and Highmoor; between Huddersfield and Halifax, and between Knaresborough and Wetherby.

In Lancashire also, Metcalf made a large extent of roads, which were of the greatest importance in opening up the resources of that county. Previous to their construction, almost the only means of communication between districts was by horse-tracks and mill-roads, of sufficient width to enable a laden horse to pass along them with a pack of goods or a sack of corn slung across its back. Metcalf's principal roads in Lancashire were those constructed by him between Bury and Blackburn, with a branch to Accrington; between Bury and Haslingden; and between Haslingden and Accrington, with a branch to Blackburn. He also made some highly important main roads connecting Yorkshire and Lancashire with each other at many parts: as, for instance, those between Skipton, Colne, and Burnley; and between Docklane Head and Ashton-under-Lyne. The roads from Ashton to Stockport and from Stockport to Mottram Langdale were also his work.

Our road-maker was also extensively employed in the same way in the counties of Cheshire and Derby; constructing the roads between Macclesfield and Chapel-le-Frith, between Whaley and Buxton, between Congleton and the Red Bull (entering Staffordshire), and in various other directions. The total mileage of the turnpike-roads thus constructed was about one hundred and eighty miles, for which Metcalf received in all about sixty-five thousand pounds. The making of these roads also involved the building of many bridges, retaining-walls, and culverts. We believe it was generally admitted of the works constructed by Metcalf that they well stood the test of time and use; and, with a degree of justifiable pride, he was afterwards accustomed to point to his bridges, when others were tumbling during floods, and boast that none of his had fallen.

This extraordinary man not only made the highways which were designed for him by other surveyors, but himself personally surveyed and laid out many of the most important roads which he constructed, in difficult and mountainous parts of Yorkshire and Lancashire. One who personally knew Metcalf thus wrote of him during his lifetime: "With the assistance only of a long staff, I have several times met this man traversing the roads, ascending steep and rugged heights, exploring valleys and investigating their several extents, forms, and situations, so as to answer his designs in the best manner. The plans which he makes, and the estimates he prepares, are done in a method peculiar to himself, and of which he cannot well convey the meaning to others. His abilities in this respect are, nevertheless, so great that he finds constant employment. Most of the roads over the Peak in Derbyshire have been altered by his directions, particularly those in the vicinity of Buxton; and he is at this time constructing a new one betwixt Wilmslow and Congleton, to open a communication with the great London road, without being obliged to pass over the mountains. I have met this blind projector while engaged in making his survey. He was alone as usual, and, amongst other conversation, I made some inquiries respecting this new road. It was really astonishing to hear with what accuracy he described its course and the nature of the different soils through which it was conducted. Having mentioned to him a boggy piece of ground it passed through, he observed that 'that was the only place he had doubts concerning, and that he was apprehensive they had, contrary to his directions, been too sparing of their materials.'" *

Metcalf's skill in constructing his roads over boggy ground was very great; and the following may be cited as an instance. When the high-road from Huddersfield to Manchester was determined on, he agreed to make it at so much a rood, though at that time the line had not been marked out. When this was done, Metcalf, to his dismay, found that the surveyor had laid it out across some deep

* 'Observations on Blindness and on the Employment of the other Senses to supply the Loss of Sight.' By Mr. Bew.—'Memoirs of the Literary and Philosophical Society of Manchester,' vol. i., pp. 172-174. Paper read 17th April, 1782.

marshy ground on Pule and Standish Commons. On this he expostulated with the trustees, alleging the much greater expense that he must necessarily incur in carrying out the work after their surveyor's plan. They told him, however, that if he succeeded in making a complete road to their satisfaction, he should not be a loser; but they pointed out that, according to their surveyor's views, it would be requisite for him to dig out the bog until he came to a solid bottom. Metcalf, on making his calculations, found that in that case he would have to dig a trench some nine feet deep and fourteen yards broad on the average, making about two hundred and ninety-four solid yards of bog in every rood, to be excavated and carried away. This, he naturally conceived, would have proved both tedious as well as costly, and, after all, the road would in wet weather have been no better than a broad ditch, and in winter liable to be blocked up with snow. He strongly represented this view to the trustees as well as the surveyor, but they were immovable. It was, therefore, necessary for him to surmount the difficulty in some other way, though he remained firm in his resolution not to adopt the plan proposed by the surveyor. After much cogitation he appeared again before the trustees, and made this proposal to them: that he should make the road across the marshes after his own plan, and then, if it should be found not to answer, he would be at the expense of making it over again after the surveyor's proposed method. This was agreed to; and as he had undertaken to make nine miles of the road within ten months, he immediately set to work with all despatch.

Nearly four hundred men were employed upon the work at six different points, and their first operation was to cut a deep ditch along either side of the intended road, and throw the excavated stuff inwards so as to raise it to a circular form. His greatest difficulty was in getting the stones laid to make the drains, there being no firm footing for a horse in the more boggy places. The Yorkshire clothiers, who passed that way to Huddersfield market—by no means a soft-spoken race—ridiculed Metcalf's proceedings, and declared that he and his men would some day have to be dragged out of the bog by the hair of their heads! Undeterred, however, by sarcasm, he persistently pursued his plan of making the road practicable for laden vehicles; but he strictly enjoined his men for the present to keep his manner of proceeding a secret.

His plan was this. He ordered heather and ling to be pulled from the adjacent ground, and after binding it together in little round bundles, which could be grasped with the hand, these bundles were placed close together in rows in the direction of the line of road, after which other similar bundles were placed transversely over them; and when all had been pressed well down, stone and gravel were led on in broad-wheeled waggons, and spread over the bundles, so as to make a firm and level way. When the first load was brought and laid on, and the horses reached the firm ground again in safety, loud cheers were set up by the persons who had assembled in the expectation of seeing both horses and waggons disappear in the bog. The whole length was finished in like manner, and it proved one of the best, and even the driest, parts of the road, standing in very little need of repair for nearly twelve years after its construction. The plan adopted by Metcalf, we need scarcely point out, was precisely similar to that afterwards adopted by George Stephenson, under like circumstances, when constructing the railway across Chat Moss. It consisted simply in a large extension of the bearing surface, by which, in fact, the road was made to float upon the surface of the bog; and the ingenuity of the expedient proved the practical shrewdness and mother-wit of the blind Metcalf, as it afterwards illustrated the promptitude as well as skill of the clear-sighted George Stephenson.

Metcalf was upwards of seventy years old before he left off road-making. He was still hale and hearty, wonderfully active for so old a man, and always full of enterprise. Occupation was absolutely necessary for his comfort, and even to the last day of his life he could not bear to be idle. While engaged on road-making in Cheshire, he brought his wife to Stockport for a time, and there she died, after thirty-nine years of happy married life. One of Metcalf's daughters became married to a person engaged in the cotton business at Stockport, and, as that trade was then very brisk, Metcalf himself commenced it in a small way. He began with six spinning-jennies and a carding-engine, to which he afterwards added looms for weaving calicoes, jeans, and velveteens. But trade was fickle, and finding that he could not sell his yarns except at a loss, he made over his jennies to his son-in-law, and again went on with his road-making. The last line which he constructed was one of the most difficult he had ever undertaken,—that between Haslingden and Accrington, with a branch road to Bury. Numerous canals being under construction at the same time, employment was abundant and wages rose, so that though he honourably fulfilled his contract, and was paid for it the sum of 3500*l*., he found himself a loser of exactly 40*l*. after two years' labour and anxiety. He completed the road in 1792, when he was seventy-five years of age, after which he retired to his farm at Spofforth, near Wetherby, where for some years longer he continued to do a little business in his old line, buying and selling hay and standing wood, and superintending the operations of his little farm. During the later years of his career he occupied himself in dictating to an amanuensis an account of the incidents in his remarkable life, and finally, in the year 1810, this strong-hearted and resolute man—his life's work over—laid down his staff and peacefully departed in the ninety-third year of his age; leaving behind him four children, twenty grand-children, and ninety great grand-children.

The roads constructed by Metcalf and others had the effect of greatly improving the communications of Yorkshire and Lancashire, and opening up those counties to the trade then flowing into them from all directions. But the administration of the highways and turnpikes being entirely local, their good or bad management depending upon the public spirit and enterprise of the gentlemen of the locality, it frequently happened that while the roads of one county were exceedingly good, those of the adjoining county were altogether execrable.

METCALF'S HOUSE AT SPOFFORTH. [By E. M. Wimperis, after Sutcliffe.]

Even in the immediate vicinity of the metropolis the Surrey roads remained comparatively unimproved. Those through the interior of Kent were wretched. When Mr. Rennie, the engineer, was engaged in surveying the Weald with a view to the cutting of a canal through it in 1802, he found the country almost destitute of practicable roads, though so near to the metropolis on the one hand and to the sea-coast on the other. The interior of the county was then comparatively untraversed, except by bands of smugglers, who kept the inhabitants in a state of constant terror. In an agricultural report on the county of Northampton as late as the year 1813, it was stated that the only way of getting along some of the main lines of road in rainy weather, was by swimming!

In the neighbourhood of the city of Lincoln the communications were little better, and there still stands upon what is called Lincoln Heath—though a heath no longer—a curious memorial of the past in the shape of Dunstan Pillar, a column seventy feet high, erected about the middle of last century in the midst of the then dreary, barren waste, for the purpose of serving as a mark to

LAND LIGHTHOUSE ON LINCOLN HEATH. [By Percival Skelton.]

wayfarers by day and a beacon to them by night.* At that time the Heath was not only uncultivated, but it was also unprovided with a road across it. When the late Lady Robert Manners visited Lincoln from her residence

* The pillar was erected by Squire Dashwood in 1751; the lantern on its summit was regularly lighted till 1788, and occasionally till 1808, when it was thrown down and never replaced. The Earl of Buckingham afterwards mounted a statue of George III. on the top.

at Bloxholm, she was accustomed to send forward a groom to examine some track, that on his return he might be able to report one that was practicable. Travellers frequently lost themselves upon this heath. Thus a family, returning from a ball at Lincoln, strayed from the track twice in one night, and they were obliged to remain there until morning. All this is now changed, and Lincoln Heath has become covered with excellent roads and thriving farmsteads. "This Dunstan Pillar," says Mr. Pusey, in his review of the agriculture of Lincolnshire, in 1843, "lighted up no longer time ago for so singular a purpose, did appear to me a striking witness of the spirit of industry which, in our own days, has reared the thriving homesteads around it, and spread a mantle of teeming vegetation to its very base. And it was certainly surprising to discover at once the finest farming I had ever seen and the only land lighthouse ever raised.* Now that the pillar has ceased to cheer the wayfarer, it may serve as a beacon to encourage other landowners in converting their dreary moors into similar scenes of thriving industry." †

When the improvement of the high roads of the country fairly set in, the progress made was very rapid. This was greatly stimulated by the important inventions of tools, machines, and engines, made towards the close of last century, the products of which—more especially of the steam-engine and spinning-machine—so largely increased

* Since the appearance of the first edition of this book, a correspondent has informed us that there is another lighthouse within 24 miles of London, not unlike that on Lincoln Heath. It is situated a little to the south-east of the Woking station of the South-Western Railway, and is popularly known as "Woking Monument." It stands on the verge of Woking Heath, which is a continuation of the vast tract of heath land which extends in one direction as far as Bagshot. The tradition among the inhabitants is, that one of the kings of England was wont to hunt in the neighbourhood, when a fire was lighted up in the beacon to guide him in case he should be belated; but the probability is, that it was erected like that on Lincoln Heath, for the guidance of ordinary wayfarers at night.

† 'Journal of the Agricultural Society of England, 1843.'

the wealth of the nation. Manufactures, commerce, and shipping, made unprecedented strides; life became more active; persons and commodities circulated more rapidly; every improvement in the internal communications being followed by an increase of ease, rapidity, and economy in locomotion. Turnpike and post roads were speedily extended all over the country, and even the rugged mountain districts of North Wales and the Scotch Highlands became as accessible as any English county. The riding postman was superseded by the smartly appointed mail-coach, performing its journeys with remarkable regularity at the average speed of ten miles an hour. Slow stage-coaches gave place to fast ones, splendidly horsed and "tooled," until travelling by road in England was pronounced almost perfect.

But all this was not enough. The roads and canals, numerous and perfect though they might be, were found altogether inadequate to the accommodation of the traffic of the country, which had increased, at a constantly accelerating ratio, with the increased application of steam power to the purposes of productive industry. At length steam itself was applied to remedy the inconveniences which it had caused; the locomotive engine was invented, and travelling by railway became generally adopted. The effect of these several improvements in the means of locomotion, has been to greatly increase the public activity, and to promote the general comfort and well-being. They have tended to bring the country and the town much closer together; and, by annihilating distance as measured by time, to make the whole kingdom as one great city. What the personal blessings of improved communication have been, no one has described so well as the witty and sensible Sydney Smith:—

"It is of some importance," he wrote, "at what period a man is born. A young man alive at this period hardly knows to what improvement of human life he has been introduced; and I would bring before his notice the changes which have taken place in England since I began to breathe the breath of life, a period amounting to over eighty years. Gas was unknown; I groped about the streets of London in the all but utter darkness of a twinkling oil lamp, under the protection of watchmen in their grand climacteric, and exposed to every species of degradation and insult. I have been nine hours in sailing from Dover to Calais, before the invention of steam. It took me nine hours to go from Taunton to Bath, before the invention of railroads; and I now go in six hours from Taunton to London! In going from Taunton to Bath, I suffered between 10,000 and 12,000 severe contusions, before stone-breaking Macadam was born. . . . . As the basket of stage-coaches in which luggage was then carried had no springs, your clothes were rubbed all to pieces; and, even in the best society, one-third of the gentlemen at least were always drunk. . . . . I paid 15*l.* in a single year for repairs of carriage-springs on the pavement of London; and I now glide without noise or fracture on wooden pavement. I can walk, by the assistance of the police, from one end of London to the other without molestation; or, if tired, get into a cheap and active cab, instead of those cottages on wheels which the hackney coaches were at the beginning of my life. . . . . Whatever miseries I suffered, there was no post to whisk my complaints for a single penny to the remotest corner of the empire; and yet, in spite of all these privations, I lived on quietly, and am now ashamed that I was not more discontented, and utterly surprised that all these changes and inventions did not occur two centuries ago.

With the history of these great improvements is also mixed up the story of human labour and genius, and of the patience and perseverance displayed in carrying them out. Probably one of the best illustrations of character in connection with the development of the inventions of the last century, is to be found in the life of Thomas Telford, the greatest and most scientific road-maker of his day, to which we proceed to direct the attention of the reader.

LIFE OF THOMAS TELFORD.

Valley of "The Unblameable Shepherd," Eskdale.

[By Percival Skelton.]

LIFE OF THOMAS TELFORD.

CHAPTER I.

ESKDALE.

THOMAS TELFORD was born in one of the most solitary nooks of the narrow valley of the Esk, in the eastern part of the county of Dumfries, in Scotland. Eskdale runs north and south, its lower end having been in former times the western march of the Scottish border. Near the entrance to the dale is a tall column erected on Langholm Hill, some twelve miles to the north of the Gretna Green station of the Caledonian Railway,—which many travellers to and from Scotland may have observed,—a monument to the late Sir John Malcolm, Governor of Bombay, one of the distinguished natives of the district. It looks far over the English border-lands, which stretch away towards the south, and marks the entrance to the mountainous parts of the dale, which lie to the north. From that point upwards the valley gradually contracts, the road winding along the river's banks, in some places high above the stream, which rushes swiftly over the rocky bed below.

A few miles upward from the lower end of Eskdale lies the little capital of the district, the town of Langholm; and there, in the market-place, stands another monument to the virtues of the Malcolm family in the statue erected to the memory of Admiral Sir Pulteney Malcolm, a distinguished naval officer. Above Langholm, the country becomes more hilly and moorland. In many places only a narrow strip of holm land by the river's side is left available for cultivation; until at length the dale contracts so much that the hills descend to the very road, and there are only to be seen their steep heathery sides sloping up towards the sky on either hand, and a narrow stream plashing and winding along the bottom of the valley among the rocks at their feet.

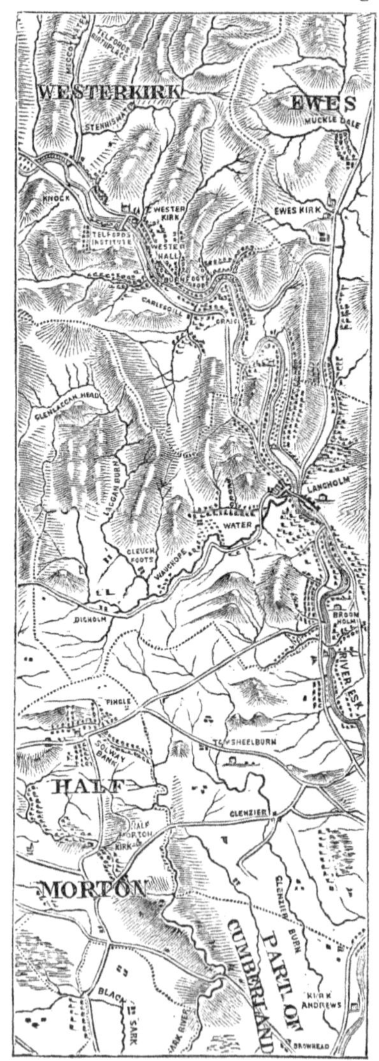

TELFORD'S NATIVE DISTRICT.

From this brief description of the character of Eskdale scenery, it may readily be supposed that the district is very thinly peopled, and that it never could have been capable of supporting a large number of inhabitants. Indeed, previous to the union of the crowns of England and Scotland, the principal branch of industry that existed in the Dale was of a lawless kind. The people living on the two sides of the border looked upon each other's cattle as their own, provided only they had the strength to "lift" them. They were, in truth, even during the time of peace, a kind of outcasts, against whom the united powers of England and Scotland were often employed. On the Scotch side of the Esk were the Johnstones and Armstrongs, and on the English the Graemes of Netherby; both clans being alike wild and lawless. It was a popular border saying that "Elliots and Armstrongs ride thieves a';" and an old historian says of the Graemes that "they were all stark moss-troopers and arrant thieves; to England as well as Scotland outlawed." The neighbouring chiefs were no better: Scott of Buccleugh, from whom the modern Duke is descended, and Scott of Harden, the ancestor of the novelist, being both renowned freebooters.

There stands at this day on the banks of the Esk, only a few miles from the English border, the ruin of an old fortalice, called Gilnockie Tower, in a situation which in point of natural beauty is scarcely equalled even in Scotland. It was the stronghold of a chief popularly known in his day as Johnnie Armstrong.* He was a mighty freebooter in the time of James V., and the terror of his name is said to have extended as far as Newcastle-upon-Tyne, between which town and his castle on the Esk he was accustomed to levy black-mail, or "protection and forbearance money," as it was called. The King, however, determining to put down by the strong hand the depredations of the march men, made a sudden expedition along the borders; and Johnnie Armstrong having been so ill-advised as to make his appearance with his followers at a place called Carlenrig, in Etterick Forest, between Hawick and Langholm, James ordered him to instant execution. Had Johnnie Armstrong, like the Scotts and Kers and Johnstones of like calling, been imprisoned beforehand, he might possibly have lived to found a British peerage; but as it was,

* Sir Walter Scott, in his notes to the 'Minstrelsy of the Scottish Border,' says that the common people of the high parts of Liddlesdale and the country adjacent to this day hold the memory of Johnnie Armstrong in very high respect.

the genius of the Armstrong dynasty was for a time extinguished, only, however, to reappear, after the lapse of a few centuries, in the person of the eminent engineer of Newcastle-upon Tyne, the inventor of the Armstrong gun.

The two centuries and a half which have elapsed since then have indeed seen extraordinary changes.* The energy which the old borderers threw into their feuds has not become extinct, but survives under more benignant aspects, exhibiting itself in efforts to enlighten, fertilize, and enrich the country which their wasteful ardour before did so much to disturb and impoverish. The heads of the Buccleugh and Elliot family now sit in the British House of Lords. The descendant of Scott of Harden has achieved a world-wide reputation as a poet and novelist; and the late Sir James Graham, the representative of the Graemes of Netherby, on the English side of the border, was one of the most venerable and respected of British statesmen. The border men, who used to make such furious raids and

* It was long before the Reformation flowed into the secluded valley of the Esk; but when it did, the energy of the Borderers displayed itself in the extreme form of their opposition to the old religion. The Eskdale people became as resolute in their covenanting as they had before been in their freebooting; the moorland fastnesses of the moss-troopers becoming the haunts of the persecuted ministers in the reign of the second James. A little above Langholm is a hill known as "Peden's View," and the well in the green hollow at its foot is still called "Peden's Well"—that place having been the haunt of Alexander Peden, the "prophet." His hiding-place was among the alder-bushes in the hollow, while from the hill-top he could look up the valley, and see whether the Johnstones of Wester Hall were coming. Quite at the head of the same valley, at a place called Craighaugh, on Eskdale Muir, one Hislop, a young covenanter, was shot by Johnstone's men, and buried where he fell; a gray slabstone still marking the place of his rest. Since that time, however, quiet has reigned in Eskdale, and its small population have gone about their daily industry from one generation to another in peace. Yet though secluded and apparently shut out by the surrounding hills from the outer world, there is not a throb of the nation's heart but pulsates along the valley; and when the author visited it some years since, he found that a wave of the great Volunteer movement had flowed into Eskdale; and the "lads of Langholm" were drilling and marching under their chief, young Mr. Malcolm of the Burnfoot, with even more zeal than in the populous towns and cities of the south.

forays, have now come to regard each other, across the imaginary line which divides them, as friends and neighbours; and they meet as competitors for victory only at agricultural meetings, where they strive to win prizes for the biggest turnips or the most effective reaping-machines; while the men who followed their Johnstone or Armstrong chiefs as prickers or hobilers to the fray have, like Telford, crossed the border with powers of road-making and bridge-building which have proved a source of increased civilization and well-being to the population of the entire United Kingdom.

The hamlet of Westerkirk, with its parish church and school, lies in a narrow part of the valley, a few miles above Langholm. Westerkirk parish is long and narrow, its boundaries being the hill-tops on either side of the dale. It is about seven miles long and two broad, with a population of about 600 persons of all ages. Yet this number is quite as much as the district is able to support, as is proved by its remaining as nearly as possible stationary from one generation to another.* But what

* The names of the families in the valley remain very nearly the same as they were three hundred years ago—the Johnstones, Littles, Scotts, and Beatties prevailing above Langholm; and the Armstrongs, Bells, Irwins, and Graemes lower down towards Canobie and Netherby. It is interesting to find that Sir David Lindesay, in his curious drama published in 'Pinkerton's Scottish Poems' (vol. ii., p. 156), gives these as among the names of the borderers some three hundred years since. One *Common Thift*, when sentenced to condign punishment, thus remembers his Border friends in his dying speech:—

"Adew! my bruther Annan thieves,
That holpit me in my mischeivis;
Adew! Grossars, Niksonis, and Bells,
Oft have we fairne owrthreuch the fells;
Adew! Robsons, Howis, and Pylis,
That in our craft hes mony wilis:
Littlis, Trumbells, and Armestranges;
Baileowes, Erewynis, and Elwandis,
Speedy of flicht, and slicht of handis;
The Scotts of Eisdale, and the Gramis,
I haf na time to tell your nameis."

Telford, or Telfer, is an old name in the same neighbourhood, commemorated in the well known border ballad of 'Jamie Telfer of the fair Dodhead.' Sir W. Scott says, in the 'Minstrelsy,' that "there is still a family of Telfers, residing near Langholm, who pretend to derive their descent from the Telfers of the Dodhead." A member of the family of "Pylis" above mentioned, is said to have migrated from Ecclefechan southward to Blackburn, and there founded the celebrated Peel family.

becomes of the natural increase of families? "They swarm off!" was the explanation given to us by a native of the valley. "If they remained at home," said he, "we should all be sunk in poverty, scrambling with each other amongst these hills for a bare living. But our peasantry have a spirit above that: they will not consent to sink; they look up; and our parish schools give them a power of making their way in the world, each man for himself. So they swarm off—some to America, some to Australia, some to India, and some, like Telford, work their way across the border and up to London."

One would scarcely have expected to find the birthplace of the builder of the Menai Bridge and other great national works in so obscure a corner of the kingdom. Possibly it may already have struck the reader with surprise, that not only were all the early engineers self-taught in their profession, but they were brought up mostly in remote country places, far from the active life of great towns and cities. But genius is of no locality, and springs alike from the farmhouse, the peasant's hut, or the herd's shieling. Strange, indeed, it is that the men who have built our bridges, docks, lighthouses, canals, and railways, should nearly all have been country-bred boys: Edwards and Brindley, the sons of small farmers; Smeaton, brought up in his father's country house at Austhorpe; Rennie, the son of a farmer and freeholder; and Stephenson, reared in a colliery village, an engine-tenter's son. But Telford, even more than any of these, was a purely country-bred boy, and was born and brought up in a valley so secluded that it could not even boast of a cluster of houses of the dimensions of a village.

Telford's father was a herd on the sheep-farm of Glendinning. The farm consists of green hills, lying along the valley of the Meggat, a little burn, which descends from the moorlands on the east, and falls into the Esk near the hamlet of Westerkirk. John Telford's cottage was little better than a shieling, consisting of four mud

TELFORD'S BIRTHPLACE.

[By R. P. Leitch.]

walls, spanned by a thatched roof. It stood upon a knoll near the lower end of a gully worn in the hillside by the torrents of many winters. The ground stretches away from it in a long sweeping slope up to the sky, and is green to the top, except where the bare grey rocks in some places crop out to the day. From the knoll may be seen miles on miles of hills up and down the valley, winding in and out, sometimes branching off into smaller glens, each with its gurgling rivulet of peaty-brown water flowing down from the mosses above. Only a narrow strip of arable land is here and there visible along the bottom of the dale, all above being sheep-pasture, moors, and rocks. At Glendinning you seem to have got almost to the world's end. There the road ceases, and above it stretch trackless moors, the solitude of which is broken only by the whimpling sound of the burns on their way to the valley below, the hum of bees gathering honey among the heather, the whirr of a blackcock on the wing, the plaintive cry of the ewes at lambing-time, or the sharp bark of the shepherd's dog gathering the flock together for the fauld.

In this cottage on the knoll Thomas Telford was born on the 9th of August, 1757, and before the year was out he was already an orphan. The shepherd, his father, died in the month of November, and was buried in Westerkirk churchyard, leaving behind him his widow and her only child altogether unprovided for. We may here mention that one of the first things which that child did, when he had grown up to manhood and could "cut a headstone," was to erect one with the following inscription, hewn and lettered by himself, over his father's grave:—

"IN MEMORY OF
JOHN TELFORD,
WHO AFTER LIVING 33 YEARS
AN UNBLAMEABLE SHEPHERD,
DIED AT GLENDINNING,
NOVEMBER, 1757,"

a simple but poetical epitaph, which Wordsworth himself might have written.

The widow had a long and hard struggle with the world before her; but she encountered it bravely. She had her boy to work for, and, destitute though she was, she had him to educate. She was helped, as the poor so often are, by those of her own condition, and there is no sense of degradation in receiving such help. One of the risks of benevolence is its tendency to lower the recipient to the condition of an alms-taker. Doles from poor's-boxes have this enfeebling effect; but a poor neighbour giving a destitute widow a help in her time of need is felt to be a friendly act, and is alike elevating to the character of both. Though misery such as is witnessed in large towns was quite unknown in the valley, there was poverty; but it was honest as well as hopeful, and none felt ashamed of it. The farmers of the dale were very primitive * in their manners and habits, and being a warm-hearted, though by no means a demonstrative race, they were kind to the widow and her fatherless boy. They took him by turns to live with them at their houses, and gave his mother occasional employment. In summer she milked the ewes and made hay, and in harvest she went a-shearing; contriving not only to live, but to be cheerful.

COTTAGE AT THE CROOKS. [By Percival Skelton.]

The house to which the widow and her son removed at the Whitsuntide following the death of her husband was at a place called The Crooks, about midway between Glendinning and Westerkirk. It was a thatched cot-house, with two ends; in one of which lived Janet Telford (more commonly known by her own name of Janet Jackson) and

* We were informed in the valley that about the time of Telford's birth there were only two tea-kettles in the whole parish of Westerkirk, one of which was in the house of Sir James Johnstone of Wester Hall, and the other at "The Burn," the residence of Mr. Pasley, grandfather of General Sir Charles Pasley.

her son Tom, and in the other her neighbour Elliot; one door being common to both.

Young Telford grew up a healthy boy, and he was so full of fun and humour that he became known in the valley by the name of "Laughing Tam." When he was old enough to herd sheep he went to live with a relative, a shepherd like his father, and he spent most of his time with him in summer on the hill-side amidst the silence of nature. In winter he lived with one or other of the neighbouring farmers. He herded their cows or ran errands, receiving for recompense his meat, a pair of stockings, and five shillings a year for clogs. These were his first wages, and as he grew older they were gradually increased.

But Tom must now be put to school, and, happily, small though the parish of Westerkirk was, it possessed the advantage of that admirable institution, the parish school. The legal provision made at an early period for the education of the people in Scotland, proved one of their greatest boons. By imparting the rudiments of knowledge to all, the parish schools of the country placed the children of the peasantry on a more equal footing with the children of the rich; and to that extent redressed the inequalities of fortune. To start a poor boy on the road of life without instruction, is like starting one on a race with his eyes bandaged or his leg tied up. Compared with the educated son of the rich man, the former has but little chance of sighting the winning post.

To our orphan boy the merely elementary teaching provided at the parish school of Westerkirk was an immense boon. To master this was the first step of the ladder he was afterwards to mount: his own industry, energy, and ability must do the rest. To school accordingly he went, still working a-field or herding cattle during the summer months. Perhaps his own "penny fee" helped to pay the teacher's hire; but it is supposed that his cousin Jackson defrayed the principal part of the expense of his instruction. It was not much that he learnt; but in acquiring the arts of reading, writing, and figures, he learnt the beginnings of a great deal.

Apart from the question of learning, there was another manifest advantage to the poor boy in mixing freely at

WESTERKIRK CHURCH AND SCHOOL. [By Percival Skelton.]

the parish school with the sons of the neighbouring farmers and proprietors. Such intercourse has an influence upon a youth's temper, manners, and tastes, which is quite as important in the education of character as the lessons of the master himself; and Telford often, in after life, referred with pleasure to the benefits which he had derived from his early school friendships. Among those to whom he was accustomed to look back with most pride, were the two elder brothers of the Malcolm family, both of whom rose to high rank in the service of their country; William Telford, a youth of great promise, a naval surgeon, who died young; and the brothers William and Andrew Little, the former of whom settled down as a farmer in Eskdale, and the latter, a surgeon, lost his eyesight when on service off the coast of Africa. Andrew Little afterwards established himself as a teacher at Langholm, where he educated, amongst others, General Sir Charles Pasley, Dr. Irving, the Custodier of the Advocate's Library at Edinburgh, and others known to fame beyond the bounds of their native valley. Well might Telford say, when an old man, full of years and honours, on sitting down to write his autobiography, "I still recollect with pride and pleasure my native parish of Westerkirk, on the banks of the Esk, where I was born."

CHAPTER II.

LANGHOLM—TELFORD LEARNS THE TRADE OF A STONEMASON.

THE time arrived when young Telford must be put to some regular calling. Was he to be a shepherd like his father and his uncle, or was he to be a farm-labourer, or put apprentice to a trade? There was not much choice; but at length it was determined to bind him to a stonemason. In Eskdale that trade was for the most part confined to the building of drystone walls, and there was very little more art employed in it than an ordinarily neat-handed labourer could manage. It was eventually decided to send the youth—and he was now a strong lad of about fifteen—to a mason at Lochmaben, a small town across the hills to the westward, where a little more building and of a better sort—such as of farm-houses, barns, and road-bridges—was carried on than in his own immediate neighbourhood. There he remained only a few months; for his master using him badly, the high-spirited youth would not brook it, and ran away, taking refuge with his mother at The Crooks, very much to her dismay.

What was now to be done with Tom? He was willing to do anything or go anywhere rather than back to his Lochmaben master. In this emergency his cousin Thomas Jackson, the factor or land-steward at Wester Hall, offered to do what he could to induce Andrew Thomson, a small mason at Langholm, to take Telford for the remainder of his apprenticeship; and to him he went accordingly. The business carried on by his new master was of a very humble sort. Telford, in his autobiography, states that most of the farmers' houses in the district then consisted of "one storey of mud walls, or rubble stones bedded in clay, and thatched with straw, rushes, or heather; the floors being of earth, and the fire in the middle, having a plastered creel chimney for the escape of the smoke; while, instead of windows, small openings in the thick mud walls admitted a scanty light." The farm-buildings were of a similarly wretched description.

The principal owner of the landed property in the neighbourhood was the Duke of Buccleugh. Shortly after the young Duke Henry succeeded to the title and estates, in 1767, he introduced considerable improvements in the farmers' houses and farm-steadings, and the peasants' dwellings, as well as in the roads throughout Eskdale. Thus a demand sprang up for masons' labour, and Telford's master had no want of regular employment for his hands. Telford profited by the experience which this increase in the building operations of the neighbourhood gave him; being employed in raising rough walls and farm enclosures, as well as in erecting bridges across rivers wherever regular roads for wheel carriages were substituted for the horse-tracks formerly in use.

During the greater part of his apprenticeship Telford lived in the little town of Langholm, taking frequent opportunities of visiting his mother at The Crooks on Saturday evenings, and accompanying her to the parish church of Westerkirk on Sundays. Langholm was then a very poor place, being no better in that respect than the district that surrounded it. It consisted chiefly of mud hovels, covered with thatch—the principal building in it being the Tolbooth, a stone and lime structure, the upper part of which was used as a justice-hall and the lower part as a gaol. There were, however, a few good houses in the little town, occupied by people of the better class, and in one of these lived an elderly lady, Miss Pasley, one of the family of the Pasleys of Craig. As the town was so small that everybody in it knew everybody else, the ruddy-cheeked, laughing mason's apprentice soon became generally known to all the townspeople, and amongst others to Miss Pasley. When she heard that he was the poor orphan boy from up the valley, the son of the hard-working widow woman, Janet Jackson, so "eident" and so industrious, her heart warmed to the mason's apprentice, and she sent for him to her house. That was a proud day for Tom; and when he called upon her, he was not more pleased with Miss Pasley's kindness than delighted at the sight of her little library of books, which contained more volumes than he had ever seen before.

Having by this time acquired a strong taste for reading, and exhausted all the little book stores of his friends, the joy of the young mason may be imagined when Miss Pasley volunteered to lend him some books from her own library. Of course, he eagerly and thankfully availed himself of the privilege; and thus, while working as an apprentice and afterwards as a journeyman, Telford gathered his first knowledge of British literature, in which he was accustomed to the close of his life to take such pleasure. He almost always had some book with him, which he would snatch a few minutes to read in the intervals of his work; and on winter evenings he occupied his spare time in poring over such volumes as came in his way, usually with no better light than the cottage fire. On one occasion Miss Pasley lent him 'Paradise Lost,' and he took the book with him to the hill-side to read. His delight was such that it fairly taxed his powers of expression to describe it. He could only say, "I read, and read, and glowred; then read, and read again." He was also a great admirer of Burns, whose writings so inflamed his mind that at the age of twenty-two, when barely out of his apprenticeship, we find the young mason actually breaking out in verse.*

* In his 'Epistle to Mr. Walter Ruddiman,' first published in 'Ruddiman's Weekly Magazine,' in 1779, occur the following lines addressed to Burns, in which Telford incidentally sketches himself at the time, and hints at his own subsequent meritorious career;—

"Nor pass the tentie curious lad,
Who o'er the ingle hangs his head,
And begs of neighbours books to read;
For hence arise
Thy country's sons, who far are spread,
Baith bold and wise."

By diligently reading all the books that he could borrow from friends and neighbours, Telford made considerable progress in his learning; and, what with his scribbling of "poetry" and various attempts at composition, he had become so good and legible a writer that he was often called upon by his less-educated acquaintances to pen letters for them to their distant friends. He was always willing to help them in this way; and, the other working people of the town making use of his services in the same manner, all the little domestic and family histories of the place soon became familiar to him. One evening a Langholm man asked Tom to write a letter for him to his son in England; and when the young scribe read over what had been written to the old man's dictation, the latter, at the end of almost every sentence, exclaimed, "Capital! capital!" and at the close he said, "Well! I declare, Tom! Werricht himsel' couldna ha' written a better!"—Wright being a well-known lawyer or "writer" in Langholm.

His apprenticeship over, Telford went on working as a journeyman at Langholm, his wages at the time being only eighteen pence a day. What was called the New Town was then in course of erection, and there are houses still pointed out in it, the walls of which Telford helped to put together. In the town are three arched door-heads of a more ornamental character than the rest, of Telford's hewing; for he was already beginning to set up his pretensions as a craftsman, and took pride in pointing to the superior handiwork which proceeded from his chisel.

About the same time, the bridge connecting the Old with the New Town was built across the Esk at Langholm, and upon that structure he was also employed. Many of the stones in it were hewn by his hand, and on several of the blocks forming the land-breast his tool-mark is still to be seen.

Not long after the bridge was finished, an unusually high flood or spate swept down the valley. The Esk was "roaring red frae bank to brae," and it was generally feared that the new brig would be carried away. Robin Hotson, the master mason, was from home at the time, and his wife, Tibby, knowing that he was bound by his contract to maintain the fabric for a period of seven years, was in a state of great alarm. She ran from one person to another, wringing her hands and sobbing, "Oh! we'll be ruined—we'll a' be ruined!" In her distress she thought of Telford, in whom she had great confidence, and called out, "Oh! where's Tammy Telfer—where's Tammy?" He was immediately sent for. It was evening, and he was soon found at the house of Miss Pasley. When he came running up, Tibby exclaimed, "Oh, Tammy! they've been on the brig, and they say its shakin'! It 'll be doon!" "Never you heed them, Tibby," said Telford, clapping her on the shoulder, "there's nae fear o' the brig. I like it a' the better that it shakes—it proves its weel put thegither." Tibby's fears, however, were not so easily allayed; and insisting that she heard the brig "rumlin," she ran up—so the neighbours afterwards used to say of her—and set her back against the parapet to hold it together. At this, it is said, "Tam hodged and leuch;" and Tibby, observing how easily he took it, at length grew more calm. It soon became clear enough that the bridge was sufficiently strong; for the flood subsided without doing it any harm, and it has stood the furious spates of nearly a century uninjured.

Telford acquired considerable general experience about the same time as a house-builder, though the structures on which he was engaged were of a humble order, being chiefly small farm-houses on the Duke of Buccleugh's estate, with the usual out-buildings. Perhaps the most important of the jobs on which he was employed was the manse of Westerkirk, where he was comparatively at home. The hamlet stands on a green hill-side, a little below the entrance to the valley of the Meggat. It consists of the kirk, the minister's manse, the parish-school, and a few cottages, every occupant of which was known to Telford. It is backed by the purple moors, up which he loved to wander in his leisure hours and read the poems of Fergusson and Burns. The river Esk gurgles along its rocky bed in the bottom of the dale, separated from the kirkyard by a steep bank, covered with natural wood; while near at hand, behind the manse, stretch the fine woods of Wester Hall, where Telford was often wont to roam. We can

VALLEY OF ESKDALE, WESTERKIRK IN THE DISTANCE. [By Percival Skelton.]

scarcely therefore wonder that, amidst such pastoral scenery, and reading such books as he did, the poetic faculty of the country mason should have become so decidedly developed. It was while working at Westerkirk manse that he sketched the first draft of his descriptive poem entitled 'Eskdale,' which was published in the 'Poetical Museum'* in 1784.

* The 'Poetical Museum,' Hawick, p. 267. 'Eskdale' was afterwards reprinted by Telford when living at Shrewsbury, when he added a few lines by way of conclusion. The poem describes very pleasantly the fine pastoral scenery of the district:—

"Deep 'mid the green sequester'd glens below,
Where murmuring streams among the alders flow,

These early poetical efforts were at least useful in stimulating his self-education. For the practice of poetical composition, while it cultivates the sentiment of beauty in thought and feeling, is probably the best of all exercises in the art of writing correctly, grammatically, and expressively. By drawing a man out of his ordinary calling, too, it often furnishes him with a power of happy thinking which may in after life become a source of the purest pleasure; and this, we believe, proved to be the case with Telford, even though he ceased in later years to pursue the special cultivation of the art.

Shortly after, when work became slack in the district, Telford undertook to do small jobs on his own account—such as the hewing of grave-stones and ornamental doorheads. He prided himself especially upon his hewing, and from the specimens of his workmanship which are still to be seen in the churchyards of Langholm and Westerkirk, he had evidently attained considerable skill. On some of these pieces of masonry the year is carved—1779, or 1780. One of the most ornamental is that set into the wall of Westerkirk church, being a monumental slab, with an inscription and moulding, surmounted by a coat of arms, to the memory of James Pasley of Craig.

He had now learnt all that his native valley could teach

Where flowery meadows down their margins spread,
And the brown hamlet lifts its humble head—
There, round his little fields, the peasant strays,
And sees his flock along the mountain graze;
And, while the gale breathes o'er his ripening grain,
And soft repeats his upland shepherd's strain,
And western suns with mellow radiance play,
And gild his straw-roof'd cottage with their ray,
Feels Nature's love his throbbing heart employ,
Nor envies towns their artificial joy."

The features of the valley are very fairly described. Its early history is then rapidly sketched; next its period of border strife, at length happily allayed by the union of the kingdoms, under which the Johnstones, Pasleys, and others, men of Eskdale, achieve honour and fame. Nor did he forget to mention Armstrong, the author of the 'Art of Preserving Health,' son of the minister of Castleton, a few miles east of Westerkirk; and Mickle, the translator of the 'Lusiad,' whose father was minister of the parish of Langholm; both of whom Telford took a natural pride in as native poets of Eskdale.

him of the art of masonry; and, bent upon self-improvement and gaining a larger experience of life, as well as knowledge of his trade, he determined to seek employment elsewhere. He accordingly left Eskdale for the first time, in 1780, and sought work in Edinburgh, where the New Town was then in course of erection on the elevated land, formerly green fields, extending along the north bank of the "Nor' Loch." A bridge had been thrown across the Loch in 1769, the stagnant pond or marsh in the hollow had been filled up, and Princes Street was rising as if by magic. Skilled masons were in great demand for the purpose of carrying out these and the numerous other architectural improvements which were in progress, and Telford had no difficulty in obtaining employment.

Our stone-mason remained at Edinburgh for about two years, during which he had the advantage of taking part in first-rate work and maintaining himself comfortably, while he devoted much of his spare time to drawing, in its application to architecture. He took the opportunity of visiting and carefully studying the fine specimens of ancient work at Holyrood House and Chapel, the Castle, Heriot's Hospital, and the numerous curious illustrations of middle age domestic architecture with which the Old Town abounds. He also made several journeys to the beautiful old chapel of Rosslyn, situated some miles to the south of Edinburgh, making careful drawings of the more important parts of that building.

When he had thus improved himself, "and studied all that was to be seen in Edinburgh, in returning to the western border," he says, "I visited the justly celebrated Abbey of Melrose." There he was charmed by the delicate and perfect workmanship still visible even in the ruins of that fine old Abbey; and with his folio filled with sketches and drawings, he made his way back to Eskdale and the humble cottage at The Crooks. But not to remain there long. He merely wished to pay a parting visit to his mother and other relatives before starting upon a longer journey. "Having acquired," he says in his Autobiography, "the rudiments of my profession, I considered that my native country afforded few opportunities of exercising it to any extent, and therefore judged it advisable (like many of my countrymen) to proceed southward, where industry might find more employment and be better remunerated."

Before setting out, he called upon all his old friends and acquaintances in the dale—the neighbouring farmers, who had befriended him and his mother when struggling with poverty—his schoolfellows, many of whom were preparing to migrate, like himself, from their native valley—and the many friends and acquaintances he had made while working as a mason in Langholm. Everybody knew that Tom was going south, and all wished him God speed. At length the leave-taking was over, and he set out for London in the year 1782, when twenty-five years old. He had, like the little river Meggat, on the banks of which he was born, floated gradually on towards the outer world: first from the nook in the valley, to Westerkirk school; then to Langholm and its little circle; and now, like the Meggat, which flows with the Esk into the ocean, he was about to be borne away into the wide world. Telford, however, had confidence in himself, and no one had fears for him. As the neighbours said, wisely wagging their heads, "Ah, he's an auld-farran chap is Tam; he'll either mak a spoon or spoil a horn; any how, he's gatten a good trade at his fingers' ends."

Telford had made all his previous journeys on foot; but this one he made on horseback. It happened that Sir James Johnstone, the laird of Wester Hall, had occasion to send a horse from Eskdale to a member of his family in London, and he had some difficulty in finding a person to take charge of it. It occurred to Mr. Jackson, the laird's factor, that this was a capital opportunity for his cousin Tom, the mason; and it was accordingly arranged that he should ride the horse to town. When a boy, he had learnt rough-riding sufficiently well for the purpose; and the better to fit him for the hardships of the road, Mr. Jackson lent him his buckskin breeches. Thus Tom set out from his native valley well mounted, with his little bundle of "traps" buckled behind him, and, after a prosperous journey, duly reached London, and delivered up the horse as he had been directed. Long after, Mr. Jackson used to tell the story of his cousin's first ride to London with great glee, and he always took care to wind up with—"but Tam forgot to send me back my breeks!"

LOWER VALLEY OF THE MEGGAT, THE CROOKS IN THE DISTANCE.

[By Percival Skelton.]

CHAPTER III.

TELFORD A WORKING MASON IN LONDON, AND FOREMAN OF MASONS AT PORTSMOUTH.

A COMMON working man, whose sole property consisted in his mallet and chisels, his leathern apron and his industry, might not seem to amount to much in "the great world of London." But, as Telford afterwards used to say, very much depends on whether the man has got a head with brains in it of the right sort upon his shoulders. In London, the weak man is simply a unit added to the vast floating crowd, and may be driven hither and thither, if he do not sink altogether; while the strong man will strike out, keep his head above water, and make a course for himself, as Telford did. There is indeed a wonderful impartiality about London. There the capable person usually finds his place. When work of importance is required, nobody cares to ask where the man who can do it best comes from, or what he has been, but what he is, and what he can do. Nor did it ever stand in Telford's way that his father had been a poor shepherd in Eskdale, and that he himself had begun his London career by working for weekly wages with a mallet and chisel.

After duly delivering up the horse, Telford proceeded to present a letter with which he had been charged by his friend Miss Pasley on leaving Langholm. It was addressed to her brother, Mr. John Pasley, an eminent London merchant, brother also of Sir Thomas Pasley, and uncle of the Malcolms. Miss Pasley requested his influence on behalf of the young mason from Eskdale, the bearer of the letter. Mr. Pasley received his countryman kindly, and furnished him with letters of introduction to Sir William

Chambers, the architect of Somerset House, then in course of erection. It was the finest architectural work in progress in the metropolis, and Telford, desirous of improving himself by experience of the best kind, wished to be employed upon it. It did not, indeed, need any influence to obtain work there, for good hewers were in demand; but our mason thought it well to make sure, and accordingly provided himself beforehand with the letter of introduction to the architect. He was employed immediately, and set to work among the hewers, receiving the usual wages for his labour.

Mr. Pasley also furnished him with a letter to Mr. Robert Adam,* another distinguished architect of the time; and Telford seems to have been much gratified by the civility which he received from him. Sir William Chambers he found haughty and reserved, probably being too much occupied to bestow attention on the Somerset House hewer, while he found Adam to be affable and communicative. "Although I derived no direct advantage from either," Telford says, "yet so powerful is manner, that the latter left the most favourable impression; while the interviews with both convinced me that my safest plan was to endeavour to advance, if by slower steps, yet by independent conduct."

There was a good deal of fine hewer's work about Somerset House, and from the first Telford aimed at taking the highest place as an artist and tradesman in that line.†

* Robert and John Adam were architects of considerable repute in their day. Among their London erections were the Adelphi Buildings, in the Strand; Lansdowne House, in Berkeley Square; Caen Wood House, near Hampstead (Lord Mansfield's); Portland Place, Regent's Park; and numerous West End streets and mansions. The screen of the Admiralty and the ornaments of Draper's Hall were also designed by them.

† Long after Telford had become famous, he was passing over Waterloo Bridge one day with a friend, when, pointing to some finely-cut stones in the corner nearest the bridge, he said: "You see those stones there; forty years since I hewed and laid them, when working on that building as a common mason."

Diligence, carefulness, and observation will always carry a man onward and upward; and before long we find that Telford had succeeded in advancing himself to the rank of a first-class mason. Judging from his letters written about this time to his friends in Eskdale, he seems to have been very cheerful and happy; and his greatest pleasure was in calling up recollections of his native valley. He was full of kind remembrances for everybody. "How is Andrew, and Sandy, and Aleck, and Davie?" he would say; and "remember me to all the folk of the nook." He seems to have made a round of the persons from Eskdale in or about London before he wrote, as his letters were full of messages from them to their friends at home; for in those days postage was dear, and as much as possible was necessarily packed within the compass of a working man's letter. In one, written after more than a year's absence, he said he envied the visit which a young surgeon of his acquaintance was about to pay to the valley; "for the meeting of long absent friends," he added, "is a pleasure to be equalled by few other enjoyments here below."

He had now been more than a year in London, during which he had acquired much practical information both in the useful and ornamental branches of architecture. Was he to go on as a working mason? or what was to be his next move? He had been quietly making his observations upon his companions, and had come to the conclusion that they very much wanted spirit, and, more than all, forethought. He found very clever workmen about him with no idea whatever beyond their week's wages. For these they would make every effort: they would work hard, exert themselves to keep their earnings up to the highest point, and very readily "strike" to secure an advance; but as for making a provision for the next week, or the next year, he thought them exceedingly thoughtless. On the Monday mornings they began "clean;" and on Saturdays their week's earnings were spent. Thus they lived from one week to another—their limited notion of "the week" seeming to bound their existence.

Telford, on the other hand, looked upon the week as only one of the storeys of a building; and upon the succession of weeks, running on through years, he thought that the complete life structure should be built up. He thus describes one of the best of his fellow-workmen at that time—the only individual he had formed an intimacy with: "He has been six years at Somerset House, and is esteemed the finest workman in London, and consequently in England. He works equally in stone and marble. He has excelled the professed carvers in cutting Corinthian capitals and other ornaments about this edifice, many of which will stand as a monument to his honour. He understands drawing thoroughly, and the master he works under looks on him as the principal support of his business. This man, whose name is Mr. Hatton, may be half a dozen years older than myself at most. He is honesty and good nature itself, and is adored by both his master and fellow-workmen. Notwithstanding his extraordinary skill and abilities, he has been working all this time as a common journeyman, contented with a few shillings a week more than the rest; but I believe your uneasy friend has kindled a spark in his breast that he never felt before." *

In fact, Telford had formed the intention of inducing this admirable fellow to join him in commencing business as builders on their own account. "There is nothing done in stone or marble," he says, "that we cannot do in the completest manner." Mr. Robert Adam, to whom the scheme was mentioned, promised his support, and said he would do all in his power to recommend them. But the great difficulty was money, which neither of them possessed; and Telford, with grief, admitting that this was an "insuperable bar," went no further with the scheme.

* Letter to Mr. Andrew Little, Langholm, dated London, July, 1783.

About this time Telford was consulted by Mr. Pulteney * respecting the alterations making in the mansion at Wester Hall, and was often with him on this business. We find him also writing down to Langholm for the prices of roofing, masonry, and timber-work, with a view to preparing estimates for a friend who was building a house in that neighbourhood. Although determined to reach the highest excellence as a manual worker, it is clear that he was already aspiring to be something more. Indeed, his steadiness, perseverance, and general ability, pointed him out as one well worthy of promotion.

How he achieved his next step we are not informed; but we find him, in July, 1784, engaged in superintending the erection of a house, after a design by Mr. Samuel Wyatt, intended for the residence of the Commissioner (now occupied by the Port Admiral) at Portsmouth Dockyard, together with a new chapel, and several buildings connected with the Yard. Telford took care to keep his eyes open to all the other works going forward in the neighbourhood, and he states that he had frequent opportunities of observing the various operations necessary in the foundation and construction of graving-docks, wharf-walls, and such like, which were among the principal occupations of his after-life.

The letters written by him from Portsmouth to his Eskdale correspondents about this time were cheerful and hopeful, like those he had sent from London. His principal grievance was that he received so few from home, but he supposed that opportunities for forwarding them by

* Mr., afterwards Sir William, Pulteney, was the second son of Sir James Johnstone, of Wester Hall, and assumed the name of Pulteney, on his marriage to Miss Pulteney, niece of the Earl of Bath and of General Pulteney, by whom he succeeded to a large fortune. He afterwards succeeded to the baronetcy of his elder brother James, who died without issue in 1797. Sir William Pulteney represented Cromarty, and afterwards Shrewsbury, where he usually resided, in seven successive Parliaments. He was a great patron of Telford's, as we shall afterwards find.

hand had not occurred, postage being so dear as scarcely then to be thought of. To tempt them to correspondence he sent copies of the poems which he still continued to compose in the leisure of his evenings: one of these was a 'Poem on Portsdown Hill.' As for himself, he was doing very well. The buildings were advancing satisfactorily; but, "above all," said he, "my proceedings are entirely approved by the Commissioners and officers here —so much so that they would sooner go by my advice than my master's, which is a dangerous point, being difficult to keep their good graces as well as his. However, I will contrive to manage it." *

The following is his own account of the manner in which he was usually occupied during the winter months while at Portsmouth Dock:—"I rise in the morning at 7 (February 1st), and will get up earlier as the days lengthen until it come to 5 o'clock. I immediately set to work to make out accounts, write on matters of business, or draw, until breakfast, which is at 9. Then I go into the Yard about 10, see that all are at their posts, and am ready to advise about any matters that may require attention. This, and going round the several works, occupies until about dinner-time, which is at 2; and after that I again go round and attend to what may be wanted. I draw till 5; then tea; and after that I write, draw, or read until half after 9; then comes supper and bed. This is my ordinary round, unless when I dine or spend an evening with a friend; but I do not make many friends, being very particular, nay, nice to a degree. My business requires a great deal of writing and drawing, and this work I always take care to keep under by reserving my time for it, and being in advance of my work rather than behind it. Then, as knowledge is my most ardent pursuit, a thousand things occur which call for investigation which would pass un-

* Letter to Andrew Little, Langholm, dated Portsmouth, July 23rd, 1784.

noticed by those who are content to trudge only in the beaten path. I am not contented unless I can give a reason for every particular method or practice which is pursued. Hence I am now very deep in chemistry. The mode of making mortar in the best way led me to inquire into the nature of lime. Having, in pursuit of this inquiry, looked into some books on chemistry, I perceived the field was boundless; but that to assign satisfactory reasons for many mechanical processes required a general knowledge of that science. I have therefore borrowed a MS. copy of Dr. Black's Lectures. I have bought his 'Experiments on Magnesia and Quicklime,' and also Fourcroy's Lectures, translated from the French by one Mr. Elliot, of Edinburgh. And I am determined to study the subject with unwearied attention until I attain some accurate knowledge of chemistry, which is of no less use in the practice of the arts than it is in that of medicine." He adds, that he continues to receive the cordial approval of the Commissioners for the manner in which he performs his duties, and says, "I take care to be so far master of the business committed to me as that none shall be able to eclipse me in that respect." * At the same time he states he is taking great delight in Freemasonry, and is about to have a lodge-room at the George Inn fitted up after his plans and under his direction. Nor does he forget to add that he has his hair powdered every day, and puts on a clean shirt three times a week.

The Eskdale mason was evidently getting on, as he deserved to do. But he was not puffed up. To his Langholm friend he averred that "he would rather have it said of him that he possessed one grain of good nature or good sense than shine the finest puppet in Christendom." "Let my mother know that I am well," he wrote to Andrew Little, "and that I will print her a letter soon." † For it

* Letter to Mr. Andrew Little, Langholm, dated Portsmouth Dockyard, Feb. 1, 1786. † Ibid.

was a practice of this good son, down to the period of his mother's death, no matter how much burdened he was with business, to set apart occasional times for the careful penning of a letter in *printed* characters, that she might the more easily be able to decipher it with her old and dimmed eyes by her cottage fireside at The Crooks. As a man's real disposition usually displays itself most strikingly in small matters—like light, which gleams the most brightly when seen through narrow chinks—it will probably be admitted that this trait, trifling though it may appear, was truly characteristic of the simple and affectionate nature of the hero of our story.

The buildings at Portsmouth were finished by the end of 1786, when Telford's duties there being at an end, and having no engagement beyond the termination of the contract, he prepared to leave, and began to look about him for other employment.

CHAPTER IV.

BECOMES SURVEYOR FOR THE COUNTY OF SALOP.

MR. PULTENEY, member for Shrewsbury, was the owner of extensive estates in that neighbourhood by virtue of his marriage with the niece of the last Earl of Bath. Having resolved to fit up the Castle there as a residence, he bethought him of the young Eskdale mason, who had, some years before, advised him as to the repairs of the Johnstone mansion at Wester Hall. Telford was soon found, and engaged to go down to Shrewsbury to superintend the necessary alterations. Their execution occupied his attention for some time, and during their progress he was so fortunate as to obtain the appointment of Surveyor of Public Works for the county of Salop, most probably through the influence of his patron. Indeed, Telford was known to be so great a favourite with Mr. Pulteney that at Shrewsbury he usually went by the name of "Young Pulteney."

Much of his attention was from this time occupied with the surveys and repairs of roads, bridges, and gaols, and the supervision of all public buildings under the control of the magistrates of the county. He was also frequently called upon by the corporation of the borough of Shrewsbury to furnish plans for the improvement of the streets and buildings of that fine old town; and many alterations were carried out under his direction during the period of his residence there.

While the Castle repairs were in course of execution, Telford was called upon by the justices to superintend the erection of a new gaol, the plans for which had already been prepared and settled. The benevolent Howard, who devoted himself with such zeal to gaol improvement, on hearing of the intentions of the magistrates, made a visit to Shrewsbury for the purpose of examining the plans; and the circumstance is thus adverted to by Telford in one of his letters to his Eskdale correspondent:—" About ten days ago I had a visit from the celebrated John Howard, Esq. I say *I*, for he was on his tour of gaols and infirmaries; and those of Shrewsbury being both under my direction, this was, of course, the cause of my being thus distinguished. I accompanied him through the infirmary and the gaol. I showed him the plans of the proposed new buildings, and had much conversation with him on both subjects. In consequence of his suggestions as to the former, I have revised and amended the plans, so as to carry out a thorough reformation; and my alterations having been approved by a general board, they have been referred to a committee to carry out. Mr. Howard also took objection to the plan of the proposed gaol, and requested me to inform the magistrates that, in his opinion, the interior courts were too small, and not sufficiently ventilated; and the magistrates, having approved his suggestions, ordered the plans to be amended accordingly. You may easily conceive how I enjoyed the conversation of this truly good man, and how much I would strive to possess his good opinion. I regard him as the guardian angel of the miserable. He travels into all parts of Europe with the sole object of doing good, merely for its own sake, and not for the sake of men's praise. To give an instance of his delicacy, and his desire to avoid public notice, I may mention that, being a Presbyterian, he attended the meeting-house of that denomination in Shrewsbury on Sunday morning, on which occasion I accompanied him; but in the afternoon he expressed a wish to attend another place of worship, his presence in the town having excited considerable curiosity, though his wish was to avoid public recognition. Nay, more, he assures me that he hates travelling, and was born to be a domestic man. He never sees his country-house but he says within himself, 'Oh! might I but rest here, and never more travel three miles from home; then should I be happy indeed!' But he has become so committed, and so pledged himself to his own conscience to carry out his great work, that he says he is doubtful whether he will ever be able to attain the desire of his heart—life at home. He never dines out, and scarcely takes time to dine at all: he says he is growing old, and has no time to lose. His manner is simplicity itself. Indeed, I have never yet met so noble a being. He is going abroad again shortly on one of his long tours of mercy."* The journey to which Telford here refers was Howard's last. In the following year he left England to return no more; and the great and good man died at Cherson, on the shores of the Black Sea, less than two years after his interview with the young engineer at Shrewsbury.

Telford writes to his Langholm friend at the same time that he is working very hard, and studying to improve himself in branches of knowledge in which he feels himself deficient. He is practising very temperate habits: for half a year past he has taken to drinking water only, avoiding all sweets, and eating no "nick-nacks." He has "sowens and milk" (oatmeal flummery) every night for his supper. His friend having asked his opinion of politics, he says he really knows nothing about them; he had been so completely engrossed by his own business that he has not had time to read even a newspaper. But, though an ignoramus in politics, he has been studying *lime*, which is more to his purpose. If his friend can give him any information about that, he will promise to read a newspaper now and then in the ensuing session of Parliament, for the purpose of forming some opinion of politics: he adds, however, "not if it interfere with my business—mind that!"

His friend told him that he proposed translating a system

* Letter to Mr. Andrew Little, Langholm, dated Shrewsbury Castle, 21st Feb., 1788.

of chemistry. "Now you know," wrote Telford, "that I am chemistry mad; and if I were near you, I would make you promise to communicate any information on the subject that you thought would be of service to your friend, especially about calcareous matters and the mode of forming the best composition for building with, as well above as below water. But not to be confined to that alone, for you must know I have a book for the pocket,* which I always carry with me, into which I have extracted the essence of Fourcroy's Lectures, Black on Quicklime, Scheele's Essays, Watson's Essays, and various points from the letters of my respected friend Dr. Irving.† So much for chemistry. But I have also crammed into it facts relating to mechanics, hydrostatics, pneumatics, and all manner of stuff, to which I keep continually adding, and it will be a charity to me if you will kindly contribute your mite."‡ He says it has been, and will continue to be, his aim to endeavour to unite those "two frequently jarring pursuits, literature and business;" and he does not see why a man should be less efficient in the latter capacity because he has well informed, stored, and humanized his mind by the cultivation of letters. There was both good sense and sound practical wisdom in this view of Telford.

While the gaol was in course of erection, after the improved plans suggested by Howard, a variety of important matters occupied the county surveyor's attention. During the summer of 1788 he says he is very much occupied, having about ten different jobs on hand: roads, bridges, streets, drainage-works, gaol, and infirmary. Yet he had

* This practice of noting down information, the result of reading and observation, was continued by Mr. Telford until the close of his life; his last pocket memorandum book, containing a large amount of valuable information on mechanical subjects—a sort of engineer's vade mecum—being printed in the appendix to the 4to. 'Life of Telford' published by his executors in 1838, pp. 663-90.

† A medical man, a native of Eskdale, of great promise, who died comparatively young.

‡ Letter to Mr. Andrew Little, Langholm.

time to write verses, copies of which he forwarded to his Eskdale correspondent, inviting his criticism. Several of these were elegiac lines, somewhat exaggerated in their praises of the deceased, though doubtless sincere. One poem was in memory of George Johnstone, Esq., a member of the Wester Hall family, and another on the death of William Telford, an Eskdale farmer's son, an intimate friend and schoolfellow of our engineer.* These, however, were but the votive offerings of private friendship, persons more immediately about him knowing nothing of his stolen pleasures in versemaking. He continued to be shy of strangers, and was very "nice," as he calls it, as to those whom he admitted to his bosom.

Two circumstances of considerable interest occurred in the course of the same year (1788), which are worthy of passing notice. The one was the fall of the church of St. Chad's, at Shrewsbury; the other was the discovery of the ruins of the Roman city of Uriconium, in the immediate neighbourhood. The church of St. Chad's was about four centuries old, and stood greatly in need of repairs. The roof let in the rain upon the congregation, and the parish vestry met to settle the plans for mending it; but they could not agree about the mode of procedure. In this emergency Telford was sent for, and requested to advise what was best to be done. After a rapid glance at

* It would occupy unnecessary space to cite these poems. The following, from the verses in memory of William Telford, relates to schoolboy days. After alluding to the lofty Fell Hills, which formed part of the sheep farm of his deceased friend s father, the poet goes on to say:—

"There 'mongst those rocks I'll form a rural seat,
And plant some ivy with its moss compleat;
I'll benches form of fragments from the stone,
Which, nicely pois'd, was by our hands o'erthrown,—
A simple frolic, but now dear to me,
Because, my Telford, 'twas performed with thee.
There, in the centre, sacred to his name,
I'll place an altar, where the lambent flame
Shall yearly rise, and every youth shall join
The willing voice, and sing the enraptured line.
But we, my friend, will often steal away
To this lone seat, and quiet pass the day;
Here oft recall the pleasing scenes we knew
In early youth, when every scene was new,
When rural happiness our moments blest,
And joys untainted rose in every breast."

the interior, which was in an exceedingly dangerous state, he said to the churchwardens, "Gentlemen, we'll consult together on the outside, if you please." He found that not only the roof but the walls of the church were in a most decayed state. It appeared that, in consequence of graves having been dug in the loose soil close to the shallow foundation of the north-west pillar of the tower, it had sunk so as to endanger the whole structure. "I discovered," says he, "that there were large fractures in the walls, on tracing which I found that the old building was in a most shattered and decrepit condition, though until then it had been scarcely noticed. Upon this I declined giving any recommendation as to the repairs of the roof unless they would come to the resolution to secure the more essential parts, as the fabric appeared to me to be in a very alarming condition. I sent in a written report to the same effect."*

The parish vestry again met, and the report was read; but the meeting exclaimed against so extensive a proposal, imputing mere motives of self-interest to the surveyor. "Popular clamour," says Telford, "overcame my report. 'These fractures,' exclaimed the vestrymen, 'have been there from time immemorial;' and there were some otherwise sensible persons, who remarked that professional men always wanted to carve out employment for themselves, and that the whole of the necessary repairs could be done at a comparatively small expense." † The vestry then called in another person, a mason of the town, and directed him to cut away the injured part of a particular pillar, in order to underbuild it. On the second evening after the commencement of the operations, the sexton was alarmed by a fall of lime-dust and mortar when he attempted to toll the great bell, on which he immediately desisted and left the church. Early next morning (on the 9th of July), while the workmen were waiting at the church door

* Letter to Mr. Andrew Little, Langholm, dated 16th July, 1788.
† Ibid.

for the key, the bell struck four, and the vibration at once brought down the tower, which overwhelmed the nave, demolishing all the pillars along the north side, and shattering the rest. "The very parts I had pointed out," says Telford, "were those which gave way, and down tumbled the tower, forming a very remarkable ruin, which astonished and surprised the vestry, and roused them from their infatuation, though they have not yet recovered from the shock." *

The other circumstance to which we have above referred was the discovery of the Roman city of Uriconium, near Wroxeter, about five miles from Shrewsbury, in the year 1788. The situation of the place is extremely beautiful, the river Severn flowing along its western margin, and forming a barrier against what were once the hostile districts of West Britain. For many centuries the dead city had slept under the irregular mounds of earth which covered it, like those of Mossul and Nineveh. Farmers raised heavy crops of turnips and grain from the surface; and they scarcely ever ploughed or harrowed the ground without turning up Roman coins or pieces of pottery. They also observed that in certain places the corn was more apt to be scorched in dry weather than in others—a sure sign to them that there were ruins underneath; and their practice, when they wished to find stones for building, was to set a mark upon the scorched places when the corn was on the ground, and after harvest to dig down, sure of finding the store of stones which they wanted for walls, cottages, or farm-houses. In fact, the place came to be regarded in the light of a quarry, rich in ready-worked materials for building purposes. A quantity of stone being wanted for the purpose of erecting a blacksmith's shop, on digging down upon one of the marked places, the labourers came upon some ancient works of a more perfect appearance than usual. Curiosity was excited—anti-

* Letter to Mr. Andrew Little, Langholm, dated 16th July, 1788.

quarians made their way to the spot—and lo! they pronounced the ruins to be neither more nor less than a Roman bath, in a remarkably perfect state of preservation. Mr. Telford was requested to apply to Mr. Pulteney, the lord of the manor, to prevent the destruction of these interesting remains, and also to permit the excavations to proceed, with a view to the buildings being completely explored. This was readily granted, and Mr. Pulteney authorised Telford himself to conduct the necessary excavations at his expense. This he promptly proceeded to do, and the result was, that an extensive hypocaust apartment was brought to light, with baths, sudatorium, dressing-room, and a number of tile pillars—all forming parts of a Roman floor—sufficiently perfect to show the manner in which the building had been constructed and used.*

Among Telford's less agreeable duties about the same time was that of keeping the felons at work. He had to devise the ways and means of employing them without risk of their escaping, which gave him much trouble and anxiety. "Really," he said, "my felons are a very troublesome family. I have had a great deal of plague from them, and I have not yet got things quite in the train that I could wish. I have had a dress made for them of white and brown cloth, in such a way that they are pyebald. They have each a light chain about one leg. Their allowance in food is a penny loaf and a halfpenny worth of cheese for breakfast; a penny loaf, a quart of soup, and half a pound of meat for dinner; and a penny loaf and a halfpenny worth of cheese for supper; so that they have meat and clothes at all events. I employ them in removing earth, serving masons or bricklayers, or in any common labouring work on which they can be employed; during which time, of course, I have them strictly watched."

* The discovery formed the subject of a paper read before the Society of Antiquaries in London on the 7th of May, 1789, published in the 'Archæologia,' together with a drawing of the remains supplied by Mr. Telford.

Much more pleasant was his first sight of Mrs. Jordan at the Shrewsbury theatre, where he seems to have been worked up to a pitch of rapturous enjoyment. She played for six nights there at the race time, during which there were various other entertainments. On the second day there was what was called an Infirmary Meeting, or an assemblage of the principal county gentlemen in the infirmary, at which, as county surveyor, Telford was present. They proceeded thence to church to hear a sermon preached for the occasion; after which there was a dinner, followed by a concert. He attended all. The sermon was preached in the new pulpit, which had just been finished after his design, in the Gothic style; and he confidentially informed his Langholm correspondent that he believed the pulpit secured greater admiration than the sermon. With the concert he was completely disappointed, and he then became convinced that he had no ear for music. Other people seemed very much pleased; but for the life of him he could make nothing of it. The only difference that he recognised between one tune and another was that there was a difference in the noise. "It was all very fine," he said, "I have no doubt; but I would not give a song of Jock Stewart * for the whole of them. The melody of sound is thrown away upon me. One look, one word of Mrs. Jordan, has more effect upon me than all the fiddlers in England. Yet I sat down and tried to be as attentive as any mortal could be. I endeavoured, if possible, to get up an interest in what was going on; but it was all of no use. I felt no emotion whatever, excepting only a strong inclination to go to sleep. It must be a defect; but it is a fact, and I cannot help it. I suppose my ignorance of the subject, and the want of musical experience in my youth, may be the cause of it." †

* An Eskdale crony. His son, Colonel Josias Stewart, rose to eminence in the East India Company's service, having been for many years Resident at Gwalior and Indore.

† Letter to Mr. Andrew Little, Langholm, dated 3rd Sept. 1788.

Telford's mother was still living in her old cottage at The Crooks. Since he had parted from her, he had written many printed letters to keep her informed of his progress; and he never wrote to any of his friends in the dale without including some message or other to his mother. Like a good and dutiful son, he had taken care out of his means to provide for her comfort in her declining years. "She has been a good mother to me," he said, "and I will try and be a good son to her." In a letter written from Shrewsbury about this time, enclosing a ten pound note, seven pounds of which were to be given to his mother, he said, "I have from time to time written William Jackson [his cousin] and told him to furnish her with whatever she wants to make her comfortable; but there may be many little things she may wish to have, and yet not like to ask him for. You will therefore agree with me that it is right she should have a little cash to dispose of in her own way. . . . I am not rich yet; but it will ease my mind to set my mother above the fear of want. That has always been my first object; and next to that, to be the *somebody* which you have always encouraged me to believe I might aspire to become. Perhaps after all there may be something in it!" *

He now seems to have occupied much of his leisure hours in miscellaneous reading. Among the numerous books which he read, he expressed the highest admiration for Sheridan's 'Life of Swift.' But his Langholm friend, who was a great politician, having invited his attention to politics, Telford's reading gradually extended in that direction. Indeed the exciting events of the French Revolution then tended to make all men more or less politicians. The capture of the Bastille by the people of Paris in 1789 passed like an electric thrill through Europe. Then followed the Declaration of Rights; after which, in the course

* Letter to Mr. Andrew Little, Langholm, dated Shrewsbury, 8th October, 1789.

of six months, all the institutions which had before existed in France were swept away, and the reign of justice was fairly inaugurated upon earth!

In the spring of 1791 the first part of Paine's 'Rights of Man' appeared, and Telford, like many others, read it, and was at once carried away by it. Only a short time before, he had admitted with truth that he knew nothing of politics; but no sooner had he read Paine than he felt completely enlightened. He now suddenly discovered how much reason he and everybody else in England had for being miserable. While residing at Portsmouth, he had quoted to his Langholm friend the lines from Cowper's 'Task,' then just published, beginning "Slaves cannot breathe in England;" but lo! Mr. Paine had filled his imagination with the idea that England was nothing but a nation of bondmen and aristocrats. To his natural mind, the kingdom had appeared to be one in which a man had pretty fair play, could think and speak, and do the thing he would, — tolerably happy, tolerably prosperous, and enjoying many blessings. He himself had felt free to labour, to prosper, and to rise from manual to head work. No one had hindered him; his personal liberty had never been interfered with; and he had freely employed his earnings as he thought proper. But now the whole thing appeared a delusion. Those rosy-cheeked old country gentlemen who came riding into Shrewsbury to quarter sessions, and were so fond of their young Scotch surveyor —occupying themselves in building bridges, maintaining infirmaries, making roads, and regulating gaols—those county magistrates and members of parliament, aristocrats all, were the very men who, according to Paine, were carrying the country headlong to ruin!

If Telford could not offer an opinion on politics before, because he "knew nothing about them," he had now no such difficulty. Had his advice been asked about the foundations of a bridge, or the security of an arch, he would have read and studied much before giving it; he would have carefully inquired into the chemical qualities of different kinds of lime—into the mechanical principles of weight and resistance, and such like; but he had no such hesitation in giving an opinion about the foundations of a constitution of more than a thousand years' growth. Here, like other young politicians, with Paine's book before him, he felt competent to pronounce a decisive judgment at once. "I am convinced," said he, writing to his Langholm friend, "that the situation of Great Britain is such, that nothing short of some signal revolution can prevent her from sinking into bankruptcy, slavery, and insignificancy." He held that the national expenditure was so enormous,* arising from the corrupt administration of the country, that it was impossible the "bloated mass" could hold together any longer; and as he could not expect that "a hundred Pulteneys," such as his employer, could be found to restore it to health, the conclusion he arrived at was that ruin was "inevitable."†

In the same letter in which these observations occur, Telford alluded to the disgraceful riots at Birmingham, in the course of which Dr. Priestley's house and library were destroyed. As the outrages were the work of the mob, Telford could not charge the aristocracy with them; but with equal injustice he laid the blame at the door of "the clergy," who had still less to do with them, winding up with the prayer, "May the Lord mend their hearts and lessen their incomes!"

Fortunately for Telford, his intercourse with the townspeople of Shrewsbury was so small that his views on these subjects were never known; and we very shortly find him

* It was then under seventeen millions sterling, or about a fourth of what it is now.

† Letter to Mr. Andrew Little, Langholm, dated 28th July, 1791. Notwithstanding the theoretical ruin of England which pressed so heavy on his mind at this time, we find Telford strongly recommending his correspondent to send any good wrights he could find in his neighbourhood to Bath, where they would be enabled to earn twenty shillings or a guinea a week at piece-work—the wages paid at Langholm for similar work being only about half those amounts.

employed by the clergy themselves in building for them a new church in the town of Bridgenorth. His patron and employer, Mr. Pulteney, however, knew of his extreme views, and the knowledge came to him quite accidentally. He found that Telford had made use of his frank to send through the post a copy of Paine's 'Rights of Man' to his Langholm correspondent,* where the pamphlet excited as much fury in the minds of some of the people of that town as it had done in that of Telford himself. The "Langholm patriots" broke out into drinking revolutionary toasts at the Cross, and so disturbed the peace of the little town that some of them were confined for six weeks in the county gaol.

Mr. Pulteney was very indignant at the liberty Telford had taken with his frank, and a rupture between them seemed likely to ensue; but the former was forgiving, and the matter went no further. It is only right to add, that as Telford grew older and wiser, he became more careful in jumping at conclusions on political topics. The events which shortly occurred in France tended in a great measure to heal his mental distresses as to the future of England. When the "liberty" won by the Parisians ran into riot, and the "Friends of Man" occupied themselves in taking off the heads of those who differed from them, he became wonderfully reconciled to the enjoyment of the substantial freedom which, after all, was secured to him by the English Constitution. At the same time, he was so much occupied

* The writer of a memoir of Telford, in the 'Encyclopedia Britannica,' says:—"Andrew Little kept a private and very small school at Langholm. Telford did not neglect to send him a copy of Paine's 'Rights of Man;' and as he was totally blind, he employed one of his scholars to read it in the evenings. Mr. Little had received an academical education before he lost his sight; and, aided by a memory of uncommon powers, he taught the classics, and particularly Greek, with much higher reputation than any other schoolmaster within a pretty extensive circuit. Two of his pupils read all the Iliad, and all or the greater part of Sophocles. After hearing a long sentence of Greek or Latin distinctly recited, he could generally construe and translate it with little or no hesitation. He was always much gratified by Telford's visits, which were not infrequent, to his native district."

in carrying out his important works, that he found but little time to devote either to political speculation or to verse-making.

While living at Shrewsbury, he had his poem of 'Eskdale' reprinted for private circulation. We have also seen several MS. verses by him, written about the same period, which do not appear ever to have been printed. One of these—the best—is entitled 'Verses to the Memory of James Thomson, author of "Liberty, a poem;"' another is a translation from Buchanan, 'On the Spheres;' and a third, written in April, 1792, is entitled 'To Robin Burns, being a postscript to some verses addressed to him on the establishment of an Agricultural Chair in Edinburgh.' It would unnecessarily occupy our space to print these effusions; and, to tell the truth, they exhibit few if any indications of poetic power. No amount of perseverance will make a poet of a man in whom the divine gift is not born. The true line of Telford's genius lay in building and engineering, in which direction we now propose to follow him.

SHREWSBURY CASTLE. [By Percival Skelton.]

CHAPTER V.

TELFORD'S FIRST EMPLOYMENT AS AN ENGINEER.

As surveyor for the county, Telford was frequently called upon by the magistrates to advise them as to the improvement of roads and the building or repair of bridges. His early experience of bridge-building in his native district now proved of much service to him, and he used often to congratulate himself, even when he had reached the highest rank in his profession, upon the circumstances which had compelled him to begin his career by working with his own hands. To be a thorough judge of work, he held that a man must himself have been practically engaged in it. "Not only," he said, "are the natural senses of seeing and feeling requisite in the examination of materials, but also the practised eye, and the hand which has had experience of the kind and qualities of stone, of lime, of iron, of timber, and even of earth, and of the effects of human ingenuity in applying and combining all these substances, are necessary for arriving at mastery in the profession; for, how can a man give judicious directions unless he possesses personal knowledge of the details requisite to effect his ultimate purpose in the best and cheapest manner? It has happened to me more than once, when taking opportunities of being useful to a young man of merit, that I have experienced opposition in taking him from his books and drawings, and placing a mallet, chisel, or trowel in his hand, till, rendered confident by the solid knowledge which experience only can bestow, he was qualified to insist on the due performance of workmanship, and to judge of merit in the lower as well as the higher departments of a profession in which no kind or degree of practical knowledge is superfluous."

The first bridge designed and built under Telford's superintendence was one of no great magnitude, across the river Severn at Montford, about four miles west of Shrewsbury. It was a stone bridge of three elliptical arches, one of 58 feet and two of 55 feet span each. The Severn at that point is deep and narrow, and its bed and banks are of alluvial earth. It was necessary to make the foundations very secure, as the river is subject to high floods; and this was effectually accomplished by means of coffer-dams. The building was substantially executed in red sandstone, and proved a very serviceable bridge, forming part of the great high road from Shrewsbury into Wales. It was finished in the year 1792.

In the same year, we find Telford engaged as an architect in preparing the designs and superintending the construction of the new parish church of St. Mary Magdalen at Bridgenorth. It stands at the end of Castle Street, near to the old ruined fortress perched upon the bold red sandstone bluff on which the upper part of the town is built. The situation of the church is very fine, and an extensive view of the beautiful vale of the Severn is obtained from it. Telford's design is by no means striking; "being," as he said, "a regular Tuscan elevation; the inside is as regularly Ionic: its only merit is simplicity and uniformity; it is surmounted by a Doric tower, which contains the bells and a clock." A graceful Gothic church would have been more appropriate to the situation, and a much finer object in the landscape; but Gothic was not then in fashion—only a mongrel mixture of many styles, without regard to either purity or gracefulness. The church, however, proved comfortable and commodious, and these were doubtless the points to which the architect paid most attention.

His completion of the church at Bridgenorth to the satisfaction of the inhabitants, brought Telford a commission, in the following year, to erect a similar edifice at Coalbrookdale. But in the mean time, to enlarge his knowledge and increase his acquaintance with the best forms of architecture, he determined to make a journey to London and through some of the principal towns of the south of England. He accordingly visited Gloucester, Worcester, and Bath, remaining several days in the last-mentioned city. He was charmed beyond expression by his journey through the manufacturing districts of Gloucestershire, more particularly by the fine scenery of the Vale of Stroud. The whole seemed to him a smiling scene of prosperous industry and middle-class comfort.

But passing out of this "Paradise," as he styled it, another stage brought him into a region the very opposite. "We stopped," says he, "at a little alehouse on the side of a rough hill to water the horses, and lo! the place was full of drunken blackguards, bellowing out 'Church and King!' A poor ragged German Jew happened to come up, whom those furious loyalists had set upon and accused of being a Frenchman in disguise. He protested that he was only a poor German who 'cut de corns,' and that all he wanted was to buy a little bread and cheese. Nothing would serve them but they must carry him before the Justice. The great brawny fellow of a landlord swore he should have nothing in his house, and, being a constable, told him that he would carry him to gaol. I interfered, and endeavoured to pacify the assailants of the poor man; when suddenly the landlord, snatching up a long knife, sliced off about a pound of raw bacon from a ham which hung overhead, and, presenting it to the Jew, swore that if he did not swallow it down at once he should not be allowed to go. The man was in a worse plight than ever. He said he was a 'poor Shoe,' and durst not eat that. In the midst of the uproar, Church and King were forgotten, and eventually I prevailed upon the landlord to accept from me as much as enabled poor little Moses to get his meal of bread and cheese; and by the time the coach started they all seemed perfectly reconciled." *

Telford was much gratified by his visit to Bath, and inspected its fine buildings with admiration. But he thought that Mr. Wood, who, he says, "created modern Bath," had left no worthy successor. In the buildings then in progress he saw clumsy designers at work, "blundering round about a meaning,"—if, indeed, there was any meaning at all in their designs, which he confessed he failed to see. From Bath he went to London by coach, making the journey in safety, "although," he says, "the collectors had been doing duty on Hounslow Heath." During his stay in London he carefully examined the principal public buildings by the light of the experience which he had gained since he last saw them. He also spent a good deal of his time in studying rare and expensive works on architecture—the use of which he could not elsewhere procure—at the libraries of the Antiquarian Society and the British Museum. There he perused the various editions of Vitruvius and Palladio, as well as Wren's 'Parentalia.' He found a rich store of ancient architectural remains in the British Museum, which he studied with great care: antiquities from Athens, Baalbec, Palmyra, and Herculaneum; "so that," he says, "what with the information I was before possessed of, and that which I have now accumulated, I think I have obtained a tolerably good general notion of architecture."

From London he proceeded to Oxford, where he carefully inspected its colleges and churches, afterwards expressing the great delight and profit which he had derived from his visit. He was entertained while there by Mr. Robertson, an eminent mathematician, then superintending the publication of an edition of the works of Archimedes. The architectural designs of buildings that most pleased him

* Letter to Mr. Andrew Little, Langholm, dated Shrewsbury, 10th March, 1793.

were those of Dr. Aldrich, Dean of Christchurch about the time of Sir Christopher Wren. He tore himself from Oxford with great regret, proceeding by Birmingham on his way home to Shrewsbury: "Birmingham," he says, "famous for its buttons and locks, its ignorance and barbarism—its prosperity increases with the corruption of taste and morals. Its nicknacks, hardware, and gilt gimcracks are proofs of the former; and its locks and bars, and the recent barbarous conduct of its populace,* are evidences of the latter." His principal object in visiting the place was to call upon a stained glass-maker respecting a window for the new church at Bridgenorth.

On his return to Shrewsbury, Telford proposed to proceed with his favourite study of architecture; but this, said he, "will probably be very slowly, as I must attend to my every day employment," namely, the superintendence of the county road and bridge repairs, and the direction of the convicts' labour. "If I keep my health, however," he added, "and have no unforeseen hindrance, it shall not be forgotten, but will be creeping on by degrees." An unforeseen circumstance, though not a hindrance, did very shortly occur, which launched Telford upon a new career, for which his unremitting study, as well as his carefully improved experience, eminently fitted him: we refer to his appointment as engineer to the Ellesmere Canal Company.

The conscientious carefulness with which Telford performed the duties entrusted to him, and the skill with which he directed the works placed under his charge, had secured the general approbation of the gentlemen of the county. His straightforward and outspoken manner had further obtained for him the friendship of many of them. At the meetings of quarter-sessions his plans had often to encounter considerable opposition, and, when called upon

* Referring to the burning of Dr. Priestley's library.

to defend them, he did so with such firmness, persuasiveness, and good temper, that he usually carried his point. "Some of the magistrates are ignorant," he wrote in 1789, "and some are obstinate: though I must say that on the whole there is a very respectable bench, and with the sensible part I believe I am on good terms." This was amply proved some four years later, when it became necessary to appoint an engineer to the Ellesmere Canal, on which occasion the magistrates, who were mainly the promoters of the undertaking, almost unanimously solicited their Surveyor to accept the office.

Indeed, Telford had become a general favourite in the county. He was cheerful and cordial in his manner, though somewhat brusque. Though now thirty-five years old, he had not lost the humorousness which had procured for him the sobriquet of "Laughing Tam." He laughed at his own jokes as well as at others. He was spoken of as jolly—a word then much more rarely as well as more choicely used than it is now. Yet he had a manly spirit, and was very jealous of his independence. All this made him none the less liked by free-minded men. Speaking of the friendly support which he had throughout received from Mr. Pulteney, he said, "His good opinion has always been a great satisfaction to me; and the more so, as it has neither been obtained nor preserved by deceit, cringing, nor flattery. On the contrary, I believe I am almost the only man that speaks out fairly to him, and who contradicts him the most. In fact, between us, we sometimes quarrel like tinkers; but I hold my ground, and when he sees I am right he quietly gives in."

Although Mr. Pulteney's influence had no doubt assisted Telford in obtaining the appointment of surveyor, it had nothing to do with the unsolicited invitation which now emanated from the county gentlemen. Telford was not even a candidate for the engineership, and had not dreamt of offering himself, so that the proposal came upon him entirely by surprise. Though he admitted he had self-confidence, he frankly confessed that he had not a sufficient amount of it to justify him in aspiring to the office of engineer to one of the most important undertakings of the day. The following is his own account of the circumstance:—

"My literary project* is at present at a stand, and may be retarded for some time to come, as I was last Monday appointed sole agent, architect, and engineer to the canal which is projected to join the Mersey, the Dee, and the Severn. It is the greatest work, I believe, now in hand in this kingdom, and will not be completed for many years to come. You will be surprised that I have not mentioned this to you before; but the fact is that I had no idea of any such appointment until an application was made to me by some of the leading gentlemen, and I was appointed, though many others had made much interest for the place. This will be a great and laborious undertaking, but the line which it opens is vast and noble; and coming as the appointment does in this honourable way, I thought it too great an opportunity to be neglected, especially as I have stipulated for, and been allowed, the privilege of carrying on my architectural profession. The work will require great labour and exertions, but it is worthy of them all."†

Telford's appointment was duly confirmed by the next general meeting of the shareholders of the Ellesmere Canal. An attempt was made to get up a party against him, but it failed. "I am fortunate," he said, "in being on good terms with most of the leading men, both of property and abilities; and on this occasion I had the decided support of the great John Wilkinson, king of the ironmasters, himself a

* The preparation of some translations from Buchanan which he had contemplated.

† Letter to Mr. Andrew Little, Langholm, dated Shrewsbury, 29th September, 1793.

host. I travelled in his carriage to the meeting, and found him much disposed to be friendly."*

The salary at which Telford was engaged was 500*l.* a year, out of which he had to pay one clerk and one confidential foreman, besides defraying his own travelling expenses. It would not appear that after making these disbursements much would remain for Telford's own labour; but in those days engineers were satisfied with comparatively small pay, and did not dream of making large fortunes.

Though Telford intended to continue his architectural business, he decided to give up his county surveyorship and other minor matters, which, he said, "give a great deal of very unpleasant labour for very little profit; in short they are like the calls of a country surgeon." One part of his forme business which he did not give up was what related o the affairs of Mr. Pulteney and Lady Bath, with whom he continued on intimate and friendly terms. He incidentally mentions in one of his letters a graceful and charming act of her Ladyship. On going into his room one day he found that, before setting out for Buxton. she had left upon his table a copy of Ferguson's 'Roman Republic,' in three quarto volumes, superbly bound and gilt.

He now looked forward with anxiety to the commencement of the canal, the execution of which would necessarily call for great exertion on his part, as well as unremitting attention and industry; "for," said he, "besides the actual labour which necessarily attends so extensive a public work, there are contentions, jealousies, and prejudices, stationed like gloomy sentinels from one extremity of the line

* John Wilkinson and his brother William were the first of the great class of ironmasters. They possessed iron forges at Bersham near Chester, at Bradley, Brimbo, Merthyr Tydvil, and other places; and became by far the largest iron manufacturers of their day. For notice of them see 'Lives of Boulton and Watt,' p. 212.

to the other. But, as I have heard my mother say that an honest man might look the Devil in the face without being afraid, so we must just trudge along in the old way."*

* Letter to Mr. Andrew Little, Langholm, dated Shrewsbury, 3rd November, 1793.

ST. MARY MAGDALEN, BRIDGENORTH.

CHAPTER VI.

THE ELLESMERE CANAL.

THE ELLESMERE CANAL consists of a series of navigations proceeding from the river Dee in the vale of Llangollen. One branch passes northward, near the towns of Ellesmere, Whitchurch, Nantwich, and the city of Chester, to Ellesmere Port on the Mersey; another, in a south-easterly direction, through the middle of Shropshire towards Shrewsbury on the Severn; and a third, in a south-westerly direction, by the town of Oswestry, to the Montgomeryshire Canal near Llanymynech; its whole extent, including the Chester Canal, incorporated with it, being about 112 miles.

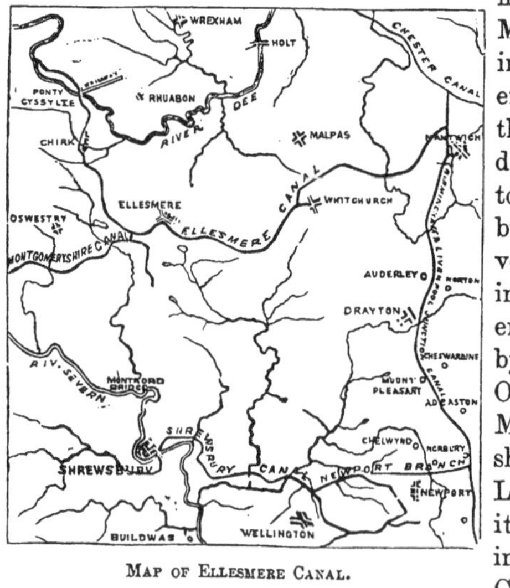

MAP OF ELLESMERE CANAL.

The success of the Duke of Bridgewater's Canal had awakened the attention of the landowners throughout England, but more especially in the districts immediately adjacent to the scene of the Duke's operations, as they saw with their own eyes the extraordinary benefits which had followed the opening up of the navigations. The resistance of the landed gentry, which many of these schemes had originally to encounter, had now completely given way, and, instead of opposing canals, they were everywhere found anxious for their construction. The navigations brought lime, coal, manure, and merchandise, almost to the farmers' doors, and provided them at the same time with ready means of conveyance for their produce to good markets. Farms in remote situations were thus placed more on an equality with those in the neighbourhood of large towns; rents rose in consequence, and the owners of land everywhere became the advocates and projectors of canals.

The dividends paid by the first companies were very high, and it was well known that the Duke's property was bringing him in immense wealth. There was, therefore, no difficulty in getting the shares in new projects readily subscribed for: indeed Mr. Telford relates that at the first meeting of the Ellesmere projectors, so eager were the public, that four times the estimated expense was subscribed without hesitation. Yet this navigation passed through a difficult country, necessarily involving very costly works; and as the district was but thinly inhabited, it did not present a very inviting prospect of dividends.* But the mania had fairly set in, and it was determined that the canal should be made. And whether the investment repaid the immediate proprietors or not, it unquestionably proved of immense advantage to the population of the districts through which it passed, and contributed to enhance the value of most of the adjoining property.

The Act authorising the construction of the canal was obtained in 1793, and Telford commenced operations very shortly after his appointment in October of the same year. His first business was to go carefully over the whole of the proposed line, and make a careful working survey, settling the levels of the different lengths, and the position of the

* The Ellesmere Canal now pays about 4 per cent. dividend.

locks, embankments, cuttings, and aqueducts. In all matters of masonry work he felt himself master of the necessary details; but having had comparatively small experience of earthwork, and none of canal-making, he determined to take the advice of Mr. William Jessop on that part of the subject; and he cordially acknowledges the obligations he was under to that eminent engineer for the kind assistance which he received from him on many occasions.

The heaviest and most important part of the undertaking was in carrying the canal through the rugged country between the rivers Dee and Ceriog, in the vale of Llangollen. From Nantwich to Whitchurch the distance is 16 miles, and the rise 132 feet, involving nineteen locks; and from thence to Ellesmere, Chirk, Pont-Cysylltau, and the river Dee, 1¾ mile above Llangollen, the distance is 38½ miles, and the rise 13 feet, involving only two locks. The latter part of the undertaking presented the greatest difficulties; as, in order to avoid the expense of constructing numerous locks, which would also involve serious delay and heavy expense in working the navigation, it became necessary to contrive means for carrying the canal on the same level from one side of the respective valleys of the Dee and the Ceriog to the other; and hence the magnificent aqueducts of Chirk and Pont-Cysylltau, characterised by Phillips as "among the boldest efforts of human invention in modern times.' *

The Chirk Aqueduct carries the canal across the valley of the Ceriog, between Chirk Castle and the village of that name. At this point the valley is above 700 feet wide; the banks are steep, with a flat alluvial meadow between them, through which the river flows. The country is finely wooded. Chirk Castle stands on an eminence on its western side, with the Welsh mountains and Glen Ceriog as a background; the whole composing a landscape of great beauty,

* 'A General History of Inland Navigation, Foreign and Domestic,' &c. By J. Phillips. Fourth edition. London, 1803.

in the centre of which Telford's aqueduct forms a highly picturesque object.

CHIRK AQUEDUCT.

[By Percival Skelton, after his orginal Drawing.]

The aqueduct consists of ten arches of 40 feet span each. The level of the water in the canal is 65 feet above the meadow, and 70 feet above the level of the river Ceriog. The proportions of this work far exceeded everything of the kind that had up to that time been attempted in England. It was a very costly structure; but Telford, like

Brindley, thought it better to incur a considerable capital outlay in maintaining the uniform level of the canal, than to raise and lower it up and down the sides of the valley by locks at a heavy expense in works, and a still greater cost in time and water. The aqueduct is a splendid specimen of the finest class of masonry, and Telford showed himself a master of his profession by the manner in which he carried out the whole details of the undertaking. The piers were carried up solid to a certain height, above which they were built hollow, with cross walls. The spandrels also, above the springing of the arches, were constructed with longitudinal walls, and left hollow.* The first stone was laid on the 17th of June, 1796, and the work was completed in the year 1801; the whole remaining in a perfect state to this day.

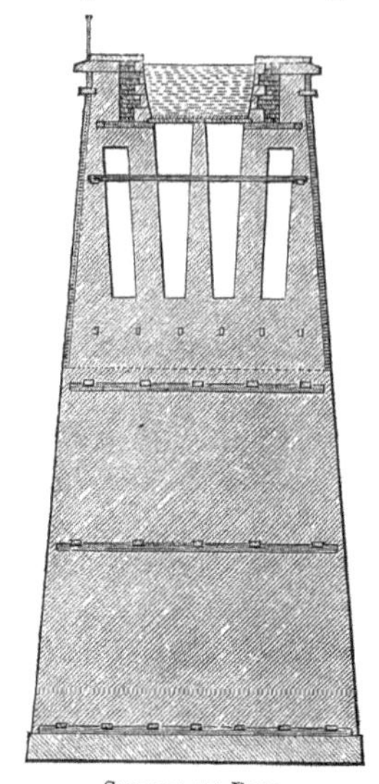

SECTION OF PIER.

* Telford himself thus modestly describes the merit of this original contrivance: "Previously to this time such canal aqueducts had been uniformly made to retain the water necessary for navigation by means of puddled earth retained by masonry; and in order to obtain sufficient breadth for this superstructure, the masonry of the piers, abutments, and arches was of massive strength; and after all this expense, and every imaginable precaution, the frosts, by swelling the moist puddle, frequently created fissures, which burst the masonry, and suffered the water to escape—nay, sometimes actually threw down the aqueducts; instances of this kind having occurred even in the works of the justly celebrated Brindley. It was evident that the increased pressure of the puddled earth was the chief cause of such failures: I therefore had recourse to the following scheme in order to avoid using it. The spandrels of the stone arches were constructed with longitudinal walls, instead of being filled in with earth (as at Kirkcudbright Bridge), and across these the canal bottom was formed by cast iron plates at each side, infixed in square stone masonry. These bottom plates had flanches on their edges, and were secured by nuts and screws at every juncture. The sides of the canal were made water-proof by ashlar masonry, backed with hard burnt

The other great aqueduct on the Ellesmere Canal, named Pont-Cysylltau, is of even greater dimensions, and a far more striking object in the landscape. Sir Walter Scott spoke of it to Southey as "the most impressive work of art he had ever seen." It is situated about four miles to the north of Chirk, at the crossing of the Dee, in the romantic vale of Llangollen. The north bank of the river is very abrupt; but on the south side the acclivity is more gradual. The lowest part of the valley in which the river runs is 127 feet beneath the water-level of the canal; and it became a question with the engineer whether the valley was to be crossed, as originally intended, by locking down one side and up the other—which would have involved seven or eight locks on each side—or by carrying it directly across by means of an aqueduct.

The execution of the proposed locks would have been very costly, and the working of them in carrying on the navigation would necessarily have involved a great waste of water, which was a serious objection, inasmuch as the supply was estimated to be no more than sufficient to provide for the unavoidable lockage and leakage of the summit level. Hence Telford was strongly in favour of an aqueduct; but, as we have already seen in the case of that at Chirk, the height of the work was such as to render it impracticable to construct it in the usual manner, upon masonry piers and arches of sufficient breadth and strength to afford room for a puddled water-way, which would have been extremely hazardous as well as expensive. He was therefore under the necessity of contriving some more safe and economical method of procedure; and he again resorted

bricks laid in Parker's cement, on the outside of which was rubble stone work, like the rest of the aqueduct. The towing path had a thin bed of clay under the gravel, and its outer edge was protected by an iron railing. The width of the water-way is 11 feet; of the masonry on each side, 5 feet 6 inches; and the depth of the water in the canal, 5 feet. By this mode of construction the quantity of masonry is much diminished and the iron bottom plate forms a continuous tie, preventing the side-walls from separation by lateral pressure of the contained water."—'Life of Telford,' p. 40.

to the practice which he had adopted in the construction of the Chirk Aqueduct, but on a much larger scale.

It will be understood that many years elapsed between the period at which Telford was appointed engineer to the Ellesmere Canal and the designing of these gigantic works. He had in the meantime been carefully gathering experience from a variety of similar undertakings on which he was employed, and bringing his observations of the strength of materials and the different forms of construction to bear upon the plans under his consideration for the great aqueducts of Chirk and Pont-Cysylltau. In 1795 he was appointed engineer to the Shrewsbury Canal, which extends from that town to the collieries and ironworks in the neighbourhood of Wrekin, crossing the rivers Roden and Tern, and Ketley Brook, after which it joins the Dorrington and Shropshire Canals. Writing to his Eskdale friend, Telford said: "Although this canal is only eighteen miles long, yet there are many important works in its

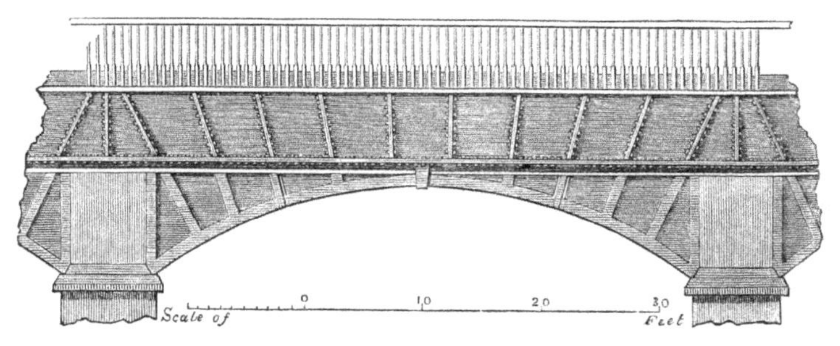

PONT-CYSYLLTAU—SIDE VIEW OF CAST IRON TROUGH.

course—several locks, a tunnel about half a mile long, and two aqueducts. For the most considerable of these last, I have just recommended an aqueduct *of iron*. It has been approved, and will be executed under my direction, upon a principle entirely new, and which I am endeavouring to establish with regard to the application of iron." *

* Letter to Mr. Andrew Little, Langholm, dated Shrewsbury, 13th March, 1795.

It was the same principle which he applied to the great aqueducts of the Ellesmere Canal now under consideration. He had a model made of part of the proposed aqueduct for Pont-Cysylltau, showing the piers, ribs, towing-path, and side railing, with a cast iron trough for the canal. The model being approved, the design was completed; the iron-work was ordered for the summit, and the masonry of the piers then proceeded. The foundation-stone was laid on the 25th July, 1795, by Richard Myddelton, Esq., of Chirk Castle, M.P., and the work was not finished until the year 1803,—thus occupying a period of nearly eight years in construction.

The aqueduct is approached on the south side by an embankment 1500 feet in length, extending from the level of the water-way in the canal until its perpendicular height at the "tip" is 97 feet; thence it is carried to the opposite side of the valley, over the river Dee, upon piers supporting nineteen arches, extending to the length of 1007 feet. The height of the piers above low water in the river is 121 feet. The lower part of each was built solid for 70 feet, all above being hollow, for the purpose of saving masonry as well as ensuring good workmanship. The outer walls of the hollow portion are only two feet thick, with cross inner walls. As each stone was exposed to inspection, and as both Telford and his confidential foreman, Matthew Davidson,* kept a vigilant eye upon the work, scamping was rendered impossible, and a first-rate piece of masonry was the result.

Upon the top of the masonry was set the cast iron trough for the canal, with its towing-path and side-rails, all accurately fitted and bolted together, forming a completely water-tight canal, with a water-way of 11 feet 10 inches, of which the towing-path, standing upon iron pillars rising

* Matthew Davidson had been Telford's fellow workman at Langholm, and was reckoned an excellent mason. He died at Inverness, where he had a situation on the Caledonian Canal.

from the bed of the canal, occupied 4 feet 8 inches, leaving a space of 7 feet 2 inches for the boat.* The whole cost of this part of the canal was 47,018*l.*, which was considered by Telford a moderate sum compared with what it must have cost if executed after the ordinary manner. The aqueduct was formally opened for traffic in 1805. "And thus," said Telford, "has been added a striking feature to the beautiful vale of Llangollen, where formerly was the fastness of Owen Glendower, but which, now cleared of its entangled woods, contains a useful line of intercourse between England and Ireland; and the water drawn from the once sacred Devon furnishes the means of distributing prosperity over the adjacent land of the Saxons."

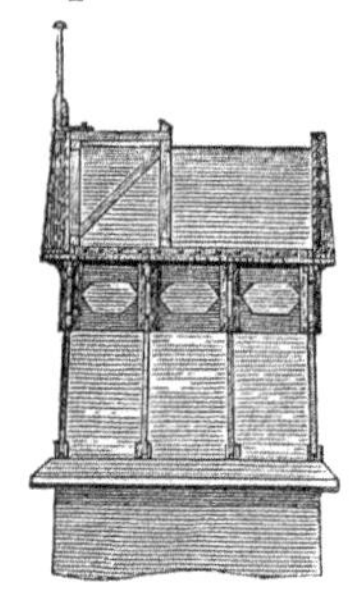

Section of Top of Pont-Cysylltau Aqueduct.

It is scarcely necessary to refer to the other works upon

* Mr. Hughes, C.E., in his 'Memoir of William Jessop,' published in 'Weale's Quarterly Papers on Engineering,' points out the bold and original idea here adopted, of constructing a water-tight trough of cast iron, in which the water of the canal was to be carried over the valleys, instead of an immense puddled trough, in accordance with the practice until that time in use; and he adds, "the immense importance of this improvement on the old practice is apt to be lost sight of at the present day by those who overlook the enormous size and strength of masonry which would have been required to support a puddled channel at the height of 120 feet." Mr. Hughes, however, claims for Mr. Jessop the merit of having suggested the employment of iron, though, in our opinion, without sufficient reason. Mr. Jessop was, no doubt, consulted by Mr. Telford on the subject; but the whole details of the design, as well as the suggestion of the use of iron (as admitted by Mr. Hughes himself), and the execution of the entire works, rested with the acting engineer. This is borne out by the report published by the Company immediately after the formal opening of the Canal in 1805, in which they state: "Having now detailed the particulars relative to the Canal, and the circumstances of the concern, the committee, in concluding their report, think it but justice due to Mr. Telford to state that the works have been planned with great skill and science, and executed with much economy and stability, doing him, as well as those employed by him, infinite credit.

(Signed) BRIDGEWATER."

PONT-CYSYLLTAU AQUEDUCT.

By PERCIVAL SKELTON

Page 162.

this canal, some of which were of considerable magnitude, though they may now seem dwarfed by comparison with the works of recent engineers. Thus, there were two difficult tunnels cut through hard rock, under the rugged ground which separates the valleys of the Dee and the Ceriog. One of these is 500 and the other 200 yards in length. To ensure a supply of water for the summit of the canal, the lake called Bala Pool was dammed up by a regulating weir, and by its means the water was drawn off at Llandisilio when required for the purposes of the navigation; the navigable feeder being six miles long, carried along the bank of the Llangollen valley. All these works were skilfully executed; and when the undertaking was finished, Mr. Telford may be said to have fairly established his reputation as an engineer of first-rate ability.

We now return to Telford's personal history during this important period of his career. He had long promised himself a visit to his dear Eskdale, and the many friends he had left there; but more especially to see his infirm mother, who had descended far into the vale of years, and longed to see her son once more before she died. He had taken constant care that she should want for nothing. *She* formed the burden of many of his letters to Andrew Little. "Your kindness in visiting and paying so much attention to her," said he, "is doing me the greatest favour which you could possibly confer upon me." He sent his friend frequent sums of money, which he requested him to lay out in providing sundry little comforts for his mother, who seems to have carried her spirit of independence so far as to have expressed reluctance to accept money even from her own son. "I must request," said he, "that you will purchase and send up what things may be likely to be wanted, either for her or the person who may be with her, as her habits of economy will prevent her from getting plenty of everything, especially as she thinks that I have to pay for it, which really hurts me more than anything else." * Though anxious to pay his intended visit, he was so occupied with one urgent matter of business and another that he feared it would be November before he could set out. He had to prepare a general statement as to the navigation affairs for a meeting of the committee; he must attend the approaching Salop quarter sessions, and after that a general meeting of the Canal Company; so that his visit must be postponed for yet another month. "Indeed," said he, "I am rather distressed at the thoughts of running down to see a kind parent in the last stage of decay, on whom I can only bestow an affectionate look, and then leave her: her mind will not be much consoled by this parting, and the impression left upon mine will be more lasting than pleasant." †

He did, however, contrive to run down to Eskdale in the following November. His mother was alive, but that was all. After doing what he could for her comfort, and providing that all her little wants were properly attended to, he hastened back to his responsible duties in connection with the Ellesmere Canal. When at Langholm, he called upon his former friends to recount with them the incidents of their youth. He was declared to be the same "canty" fellow as ever, and, though he had risen greatly in the world, he was "not a bit set up." He found one of his old fellow workmen, Frank Beattie, become the principal innkeeper of the place. "What have you made of your mell and chisels?" asked Telford. "Oh!" replied Beattie, "they are all dispersed—perhaps lost." "I have taken better care of mine," said Telford; "I have them all locked up in a room at Shrewsbury, as well as my old working clothes and leather apron: you know one can never tell what may happen."

He was surprised, as most people are who visit the scenes of their youth after a long absence, to see into what small

* Letter to Mr. Andrew Little, Langholm, dated Shrewsbury, 16th Sept., 1794.

† Ibid.

dimensions Langholm had shrunk. That High Street, which before had seemed so big, and that frowning gaol and court-house in the Market Place, were now comparatively paltry to eyes that had been familiar with Shrewsbury, Portsmouth, and London. But he was charmed, as ever, with the sight of the heather hills and the narrow winding valley—

> "Where deep and low the hamlets lie
> Beneath their little patch of sky,
> And little lot of stars."

On his return southward, he was again delighted by the sight of old Gilnockie Castle and the surrounding scenery. As he afterwards wrote to his friend Little, "Broomholm was in all his glory." Probably one of the results of this visit was the revision of the poem of 'Eskdale,' which he undertook in the course of the following spring, putting in some fresh touches and adding many new lines, whereby the effect of the whole was considerably improved. He had the poem printed privately, merely for distribution amongst friends; "being careful," as he said, that "no copies should be smuggled and sold."

Later in the year we find him, on his way to London on business, sparing a day or two for the purpose of visiting the Duke of Buckingham's palace and treasures of art at Stowe; afterwards writing out an eight page description of it for the perusal of his friends at Langholm. At another time, when engaged upon the viaduct at Pont-Cysylltau, he snatched a few day's leisure to run through North Wales, of which he afterwards gave a glowing account to his correspondent. He passed by Cader Idris, Snowdon, and Penmaen Mawr. "Parts of the country we passed through," he says, "very much resemble the lofty green hills and woody vales of Eskdale. In other parts the magnificent boldness of the mountains, the torrents, lakes, and waterfalls, give a special character to the scenery, unlike everything of the kind I had before seen. The vale of Llanrwst is peculiarly beautiful and fertile. In this vale is the celebrated bridge of Inigo Jones; but what is a much more delightful circumstance, the inhabitants of the vale are the most beautiful race of people I have ever beheld; and I am much astonished that this never seems to have struck the Welsh tourists. The vale of Llangollen is very fine, and not the least interesting object in it, I can assure you, is Davidson's famous aqueduct [Pont-Cysylltau], which is already reckoned among the wonders of Wales. Your old acquaintance thinks nothing of having three or four carriages at his door at a time." *

It seems that, besides attending to the construction of the works, Telford had to organise the conduct of the navigation at those points at which the canal was open for traffic. By the middle of 1797 he states that twenty miles were in working condition, along which coal and lime were conveyed in considerable quantities, to the profit of the Company and the benefit of the public; the price of these articles having already in some places been reduced twenty-five, and in others as much as fifty, per cent. "The canal affairs," he says in one of his letters, "have required a good deal of exertion, though we are on the whole doing well. But, besides carrying on the works, it is now necessary to bestow considerable attention on the creating and guiding of a trade upon those portions which are executed. This involves various considerations, and many contending and sometimes clashing interests. In short, it is the working of a great machine: in the first place, to draw money out of the pockets of a numerous proprietary to make an expensive canal, and then to make the money return into their pockets by the creation of a business upon that canal."

But, as if all this business were not enough, he was occupied at the same time in writing a book upon the

* Letter to Mr. Andrew Little, Langholm, dated Salop, 20th Aug., 1797.

subject of Mills. In the year 1796 he had undertaken to draw up a paper on this topic for the Board of Agriculture, and by degrees it had grown into a large quarto volume, illustrated by upwards of thirty plates. He was also reading extensively in his few leisure moments; and among the solid works which he perused we find him mentioning Robertson's 'Disquisitions on Ancient India,' Stewart's 'Philosophy of the Human Mind,' and Alison's 'Principles of Taste.' As a relief from these graver studies, he seems, above all things, to have taken peculiar pleasure in occasionally throwing off a bit of poetry. Thus, when laid up at an hotel in Chester by a blow on his leg, which disabled him for some weeks, he employed part of his time in writing his 'Verses on hearing of the Death of Robert Burns.' On another occasion, when on his way to London, and detained for a night at Stratford-on-Avon, he occupied the evening at his inn in composing some stanzas, entitled 'An Address to the River Avon.' And when on his way back to Shrewsbury, while resting for the night at Bridgenorth, he amused himself with revising and copying out the verses for the perusal of Andrew Little. "There are worse employments," he said, "when one has an hour to spare from business;" and he asked his friend's opinion of the composition. It seems to have been no more favourable than the verses deserved; for, in his next letter, Telford says, "I think your observations respecting the verses to the Avon are correct. It is but seldom I have time to versify; but it is to me something like what a fiddle is to others. I apply to it in order to relieve my mind, after being much fatigued with close attention to business."

It is very pleasant to see the engineer relaxing himself in this way, and submitting cheerfully to unfavourable criticism, which is so trying to even the best of tempers. The time, however, thus taken from his regular work was not loss, but gain. Taking the character of his occupation into account, it was probably the best kind of relaxation he could have indulged in. With his head full of bridges and viaducts, he thus kept his heart open to the influences of beauty in life and nature; and, at all events, the writing of verses, indifferent though they might have been, proved of this value to him—that it cultivated in him the art of writing better prose.

CHAPTER VII.

IRON AND OTHER BRIDGES.

SHREWSBURY being situated in the immediate neighbourhood of the Black Country, of which coal and iron are the principal products, Telford's attention was naturally directed, at a very early period, to the employment of cast iron in bridge-building. The strength as well as lightness of a bridge of this material, compared with one of stone and lime, is of great moment where headway is of importance, or the difficulties of defective foundations have to be encountered. The metal can be moulded in such precise forms and so accurately fitted together as to give to the arching the greatest possible rigidity; while it defies the destructive influences of time and atmospheric corrosion with nearly as much certainty as stone itself.

The Italians and French, who took the lead in engineering down almost to the end of last century, early detected the value of this material, and made several attempts to introduce it in bridge-building; but their efforts proved unsuccessful, chiefly because of the inability of the early founders to cast large masses of iron, and also because the metal was then more expensive than either stone or timber. The first actual attempt to build a cast iron bridge was made at Lyons in 1755, and it proceeded so far that one of the arches was put together in the builder's yard; but the project was abandoned as too costly, and timber was eventually used.

It was reserved for English manufacturers to triumph over the difficulties which had baffled the foreign iron-founders. Shortly after the above ineffectual attempt had been made, the construction of a bridge over the Severn near Broseley formed the subject of discussion among the adjoining owners. There had been a great increase in the coal, iron, brick, and pottery trades of the neighbourhood; and the old ferry between the opposite banks of the river was found altogether inadequate for the accommodation of the traffic. The necessity for a bridge had long been felt, and the project of constructing one was actively

THE FIRST IRON BRIDGE, COALBROOKDALE. [By E. M. Wimperis.]

taken up in 1776 by Mr. Abraham Darby, the principal owner of the extensive iron works at Coalbrookdale. Mr. Pritchard, a Shrewsbury architect, prepared the design of a stone bridge of one arch, in which he proposed to introduce a key-stone of cast iron, occupying only a few feet at the crown of the arch. This plan was, however, given up as unsuitable; and another, with the entire arch of cast iron, was designed under the superintendence of Mr. Darby. The castings were made in the works at Coalbrookdale, and the bridge was erected at a point where the banks were of considerable height on both sides of the river. It was opened for traffic in 1779, and continues a most serviceable structure to this day, giving the name to the town of Ironbridge, which has sprung up in its immediate vicinity. The bridge consists of one semicircular arch, of 100 feet span, each of the great ribs consisting of two pieces only. Mr. Robert Stephenson has said of the structure—"If we consider that the manipulation of cast iron was then completely in its infancy, a bridge of such dimensions was doubtless a bold as well as an original undertaking, and the efficiency of the details is worthy of the boldness of the conception."*

It is a curious circumstance that the next projector of an iron bridge—and that of a very bold design—was the celebrated, or rather the notorious, Tom Paine, whose political writings Telford had so much admired. The son of a decent Quaker of Thetford, who trained him to his own trade of a staymaker, Paine seems early to have contracted a dislike for the sect to which his father belonged. Arrived at manhood, he gave up staymaking to embrace the wild life of a privateersman, and served in two successive adventures. Leaving the sea, he became an exciseman, but retained his commission for only a year. Then he became an usher in a school, during which he studied mechanics and mathematics. Again appointed an exciseman, he was stationed at Lewes in Sussex, where he wrote poetry and acquired some local celebrity as a writer. He was accordingly selected by his brother excisemen to prepare their petition to Government for an increase of

* 'Encyclopedia Britannica,' 8th ed. Art. "Iron Bridges."

pay,*—the document which he drew up procuring him introductions to Goldsmith and Franklin, and dismissal from his post. Franklin persuaded him to go to America; and there the quondam staymaker, privateersman, usher, poet, and exciseman, took an active part in the revolutionary discussions of the time, besides holding the important office of Secretary to the Committee for Foreign Affairs. Paine afterwards settled for a time at Philadelphia, where he occupied himself with the study of mechanical philosophy, electricity, mineralogy, and the use of iron in bridge-building. In 1787, when a bridge over the Schuylkill was proposed, without any river piers, as the stream was apt to be choked with ice in the spring freshets, Paine boldly offered to build an iron bridge with a single arch of 400 feet span. In the course of the same year, he submitted his design of the proposed bridge to the Academy of Sciences at Paris; he also sent a copy of his plan to Sir Joseph Banks for submission to the Royal Society; and, encouraged by the favourable opinions of scientific men, he proceeded to Rotherham, in Yorkshire, to have his bridge cast.† An American gentleman, named Whiteside, having advanced money to Paine on security of his property in the States, to enable the bridge to be completed, the castings were duly made, and shipped off to London, where they were put together and exhibited to the public on a bowling-green at Paddington. The bridge was there visited by a large number of persons, and was considered to be a highly creditable work. Suddenly Paine's attention was withdrawn from its further prosecution by the publication of Mr. Burke's celebrated 'Thoughts on the French Revolution,' which he undertook to answer. Whiteside having in the meantime become bankrupt, Paine was arrested by

* According to the statement made in the petition drawn by Paine, excise officers were then (1772) paid only 1*s.* 9½*d.* a day.

† In England, Paine took out a patent for his Iron Bridge in 1788.—Specification of Patents (old law) No. 1667.

his assignees, but was liberated by the assistance of two other Americans, who became bound for him. Paine, however, was by this time carried away by the fervour of the French Revolution, having become a member of the National Convention, as representative for Calais. The "Friends of Man," whose cause he had espoused, treated him scurvily, imprisoning him in the Luxembourg, where he lay for eleven months. Escaped to America, we find him in 1803 presenting to the American Congress a memoir on the construction of Iron Bridges, accompanied by several models. It does not appear, however, that Paine ever succeeded in erecting an iron bridge. He was a restless, speculative, unhappy being; and it would have been well for his memory if, instead of penning shallow infidelity, he had devoted himself to his original idea of improving the communications of his adopted country. In the meantime, however, the bridge exhibited at Paddington had produced important results. The manufacturers agreed to take it back as part of their debt, and the materials were afterwards used in the construction of the noble bridge over the Wear at Sunderland, which was erected in 1796.

The project of constructing a bridge at this place, where the rocky banks of the Wear rise to a great height on both sides of the river, is due to Rowland Burdon, Esq., of Castle Eden, under whom Mr. T. Wilson served as engineer in carrying out his design. The details differed in several important respects from the proposed bridge of Paine, Mr. Burdon introducing several new and original features, more particularly as regarded the framed iron panels radiating towards the centre in the form of voussoirs, for the purpose of resisting compression. Mr. Phipps, C.E., in a report prepared by him at the instance of the late Robert Stephenson, under whose superintendence the bridge was recently repaired, observes, with respect to the original design,—"We should probably make a fair division of the honour connected with this unique bridge, by conceding to Burdon all that belongs to a careful elaboration and improvement upon the designs of another, to the boldness of taking upon himself the great responsibility of applying this idea at once on so magnificent a scale, and to his liberality and public spirit in furnishing the requisite funds [to the amount of 22,000*l.*]; but we must not deny

WEAR BRIDGE, AT SUNDERLAND. [By Percival Skelton.]

to Paine the credit of conceiving the construction of iron bridges of far larger span than had been made before his time, or of the important examples both as models and large constructions which he caused to be made and publicly exhibited. In whatever shares the merit of this great work may be apportioned, it must be admitted to be one of the earliest and greatest triumphs of the art of bridge construction." Its span exceeded that of any arch then known, being 236 feet, with a rise of 34 feet, the springing commencing at 95 feet above the bed of the river; and its height was such as to allow vessels of 300 tons burden to sail underneath without striking their masts. Mr. Stephenson characterised the bridge as "a structure which, as regards its proportions and the small quantity of material employed in its construction, will probably remain unrivalled."

The same year in which Burdon's Bridge was erected at Sunderland, Telford was building his first iron bridge over the Severn at Buildwas, at a point about midway between Shrewsbury and Bridgenorth. An unusually high flood having swept away the old bridge in the year 1795, he was called upon, as surveyor for the county, to supply the plan of a new one. Having carefully examined the bridge at Coalbrookdale, and appreciated its remarkable merits, he determined to build the proposed bridge at Buildwas of iron; and as the waters came down with great suddenness from the Welsh mountains, he further resolved to construct it of only one arch, so as to afford the largest possible water-way.

He had some difficulty in inducing the Coalbrookdale iron-masters, who undertook the casting of the girders, to depart from the plan of the earlier structure; but he persisted in his design, which was eventually carried out. It consisted of a single arch of 130 feet span, the segment of a very large circle, calculated to resist the tendency of the abutments to slide inwards, which had been a defect of the Coalbrookdale bridge; the flat arch being itself sustained and strengthened by an outer ribbed one on each side, springing lower than the former and also rising higher, somewhat after the manner of timber-trussing. Although the span of the new bridge was 30 feet wider than the Coalbrookdale bridge, it contained less than half the quantity of iron; Buildwas bridge containing 173, whereas the other contained 378 tons. The new structure was, besides, extremely elegant in form; and when the centres were struck, the arch and abutments stood perfectly firm, and have remained so to this day. But the ingenious design of this bridge will be better explained by the following representation than by any description in words.*

BUILDWAS BRIDGE. [By Percival Skelton.]

The bridge at Buildwas, however, was not Telford's first employment of iron in bridge-building; for, the year before its erection, we find him writing to his friend at Langholm that he had recommended an iron aqueduct for the Shrewsbury Canal, "on a principle entirely new," and which he was "endeavouring to establish with regard to

* The following are further details: "Each of the main ribs of the flat arch consists of three pieces, and at each junction they are secured by a grated plate, which connects all the parallel ribs together into one frame. The back of each abutment is in a wedge-shape, so as to throw off laterally much of the pressure of the earth. Under the bridge is a towing path on each side of the river. The bridge was cast in an admirable manner by the Coalbrookdale iron-masters in the year 1796, under contract with the county magistrates. The total cost was 6034*l*. 13*s*. 3*d*."

the application of iron."* This iron aqueduct had been cast and fixed; and it was found to effect so great a saving in masonry and earthwork, that he was afterwards induced to apply the same principle, as we have already seen, in different forms, in the magnificent aqueducts of Chirk and Pont-Cysylltau.

The uses of cast iron in canal construction became more obvious with every year's successive experience; and Telford was accustomed to introduce it in many cases where formerly only timber or stone had been used. On the Ellesmere, and afterwards on the Caledonial Canal, he adopted cast iron lock-gates, which were found to answer well, being more durable than timber, and not liable like it to shrink and expand with alternate dryness and wet. The turnbridges which he applied to his canals, in place of the old drawbridges, were also of cast iron; and in some cases even the locks were of the same material. Thus, on a part of the Ellesmere Canal opposite Beeston Castle, in Cheshire, where a couple of locks, together rising 17 feet, having been built on a stratum of quicksand, were repeatedly undermined, the idea of constructing the entire locks of cast iron was suggested; and this unusual application of the new material was accomplished with entirely satisfactory results.

But Telford's principal employment of cast iron was in the construction of road bridges, in which he proved himself a master. His experience in these structures had become very extensive. During the time that he held the office of surveyor to the county of Salop, he erected no fewer than forty-two, five of which were of iron. Indeed, his success in iron bridge-building so much emboldened him, that in 1801, when Old London Bridge had become so rickety and inconvenient that it was found necessary to take steps to rebuild or remove it, he proposed the daring

* Letter to Mr. Andrew Little, Langholm, dated Shrewsbury, 18th March, 1795.

plan of a cast iron bridge of a single arch of not less than 600 feet span, the segment of a circle 1450 feet in diameter. In preparing this design we find that he was associated with a Mr. Douglas, to whom many allusions are made in his private letters.* The design of this bridge seems to have arisen out of a larger project for the improvement of the port of London. In a private letter of Telford's, dated the 13th May, 1800, he says:—

"I have twice attended the Select Committee on the Port of London, Lord Hawkesbury, Chairman. The subject has now been agitated for four years, and might have been so for many more, if Mr. Pitt had not taken the business out of the hands of the General Committee, and got it referred to a Select Committee. Last year they recommended that a system of docks should be formed in a large bend of the river opposite Greenwich, called the Isle of Dogs, with a canal across the neck of the bend. This part of the contemplated improvements is already commenced, and is proceeding as rapidly as the nature of the work will admit. It will contain ship docks for large vessels, such as East and West Indiamen, whose draught of water is considerable.

"There are now two other propositions under consideration. One is to form another system of docks at Wapping, and the other to take down London Bridge, rebuild it of such dimensions as to admit of ships of 200 tons passing under it, and form a new pool for ships of such burden between London and Blackfriars Bridges, with a set of regular wharves on each side of the river. This is

---

* Douglas was first mentioned to Telford, in a letter from Mr. Pasley, as a young man, a native of Bigholmes, Eskdale, who had, after serving his time there as a mechanic, emigrated to America, where he showed such proofs of mechanical genius that he attracted the notice of Mr. Liston, the British Minister, who paid his expenses home to England, that his services might not be lost to his country, and at the same time gave him a letter of introduction to the Society of Arts in London. Telford, in a letter to Andrew Little, dated 4th December, 1797, expressed a desire "to know more of this Eskdale Archimedes." Shortly after, we find Douglas mentioned as having invented a brick machine, a shearing-machine, and a ball for destroying the rigging of ships; for the two former of which he secured patents. He afterwards settled in France, where he introduced machinery for the improved manufacture of woollen cloth; and being patronised by the Government, he succeeded in realising considerable wealth, which, however, he did not live to enjoy.

with the view of saving lighterage and plunderage, and bringing the great mass of commerce so much nearer to the heart of the City. This last part of the plan has been taken up in a great measure from some statements I made while in London last year, and I have been called before the Committee to explain. I had previously prepared a set of plans and estimates for the purpose of showing how the idea might be carried out; and thus a considerable degree of interest has been excited on the subject. It is as yet, however, very uncertain how far the plans will be carried out. It is certainly a matter of great national importance to render the port of London as perfect as possible." *

Later in the same year he writes that his plans and propositions have been approved and recommended to be carried out, and he expects to have the execution of them. "If they will provide the ways and means," says he, "and give me elbow-room, I see my way as plainly as mending the brig at the auld burn." In November, 1801, he states that his view of London Bridge, as proposed by him, has been published, and much admired. On the 14th of April, 1802, he writes, "I have got into mighty favour with the Royal folks. I have received notes written by order of the King, the Prince of Wales, Duke of York, and Duke of Kent, about the bridge print, and in future it is to be dedicated to the King."

The bridge in question was one of the boldest of Telford's designs. He proposed by his one arch to provide a clear headway of 65 feet above high water. The arch was to consist of seven cast iron ribs, in segments as large as possible, and they were to be connected by diagonal cross-bracing, disposed in such a manner that any part of the ribs and braces could be taken out and replaced without injury to the stability of the bridge or interruption to the traffic over it. The roadway was to be 90 feet wide at the abutments and 45 feet in the centre; the width of the arch being gradually contracted towards the crown in order

* Letter to Mr. Andrew Little, Langholm, dated London, 13th May, 1800.

to lighten the weight of the structure. The bridge was to contain 6500 tons of iron, and the cost of the whole was to be 262,289*l*.

The originality of the design was greatly admired, though there were many who received with incredulity the proposal to bridge the Thames by a single arch, and it was sarcastically said of Telford that he might as well

TELFORD'S PROPOSED ONE-ARCHED BRIDGE OVER THE THAMES.

think of "setting the Thames on fire." Before any outlay was incurred in building the bridge, the design was submitted to the consideration of the most eminent scientific and practical men of the day; after which evidence was taken at great length before a Select Committee which sat on the subject. Among those examined on the occasion were the venerable James Watt of Birmingham, Mr. John Rennie, Professor Hutton of Woolwich, Professors Playfair and Robison of Edinburgh, Mr. Jessop, Mr. Southern, and Dr. Maskelyne. Their evidence will still be found interesting as indicating the state at which constructive science had at that time arrived in England.*

* The evidence is fairly set forth in 'Cresy's Encyclopedia of Civil Engineering,' p. 475.

There was a considerable diversity of opinion among the witnesses, as might have been expected; for experience was as yet very limited as to the resistance of cast iron to extension and compression. Some of them anticipated immense difficulty in casting pieces of metal of the necessary size and exactness, so as to secure that the radiated joints should be all straight and bearing. Others laid down certain ingenious theories of the arch, which did not quite square with the plan proposed by the engineer. But, as was candidly observed by Professor Playfair in concluding his report—"It is not from theoretical men that the most valuable information in such a case as the present is to be expected. When a mechanical arrangement becomes in a certain degree complicated, it baffles the efforts of the geometer, and refuses to submit to even the most approved methods of investigation. This holds good particularly of bridges, where the principles of mechanics, aided by all the resources of the higher geometry, have not yet gone further than to determine the equilibrium of a set of smooth wedges acting on one another by pressure only, and in such circumstances as, except in a philosophical experiment, can hardly ever be realised. It is, therefore, from men educated in the school of daily practice and experience, and who to a knowledge of general principles have added, from the habits of their profession, a certain feeling of the justness or insufficiency of any mechanical contrivance, that the soundest opinions on a matter of this kind can be obtained."

It would appear that the Committee came to the general conclusion that the construction of the proposed bridge was practicable and safe; for the river was contracted to the requisite width, and the preliminary works were actually begun. Mr. Stephenson says the design was eventually abandoned, owing more immediately to the difficulty of constructing the approaches with such a head way, which would have involved the formation of extensive inclined planes from the adjoining streets, and thereby led to serious inconvenience, and the depreciation of much valuable property on both sides of the river.*

Telford's noble design of his great iron bridge over the Thames, together with his proposed embankment of the river, being thus definitely abandoned, he fell back upon his ordinary business as an architect and engineer, in the course of which he designed and erected several stone bridges of considerable magnitude and importance.

In the spring of 1795, after a long continued fall of snow, a sudden thaw raised a heavy flood in the Severn, which carried away many bridges—amongst others one at Bewdley, in Worcestershire,—when Telford was called upon to supply a design for a new structure. At the same time, he was required to furnish a plan for a new bridge near the town of Bridgenorth; "in short," he wrote to his friend, "I have been at it night and day." So uniform a success had heretofore attended the execution of his designs, that his reputation as a bridge-builder was universally acknowledged. "Last week," he says, "Davidson and I struck the centre of an arch of 76 feet span, and this is the third which has been thrown this summer, none of which have shrunk a quarter of an inch."

Bewdley Bridge is a handsome and substantial piece of masonry. The streets on either side of it being on low ground, land arches were provided at both ends for the passage of the flood waters; and as the Severn was navigable at the point crossed, it was considered necessary to allow considerably greater width in the river arches than had been the case in the former structure. The arches were three in number—one of 60 feet span and two of 52 feet, the land arches being of 9 feet span. The works were proceeded with and the bridge was completed during the summer of 1798, Telford writing to his friend in December of that year—"We have had a remarkably dry summer

* Article on Iron Bridges, in the 'Encyclopedia Britannica,' Edinburgh, 1857.

and autumn; after that an early fall of snow and some frost, followed by rain. The drought of the summer was unfavourable to our canal working; but it has enabled us to raise Bewdley Bridge as if by enchantment. We have thus built a magnificent bridge over the Severn in one

BEWDLEY BRIDGE. [By Percival Skelton.]

season, which is no contemptible work for John Simpson* and your humble servant, amidst so many other great undertakings. John Simpson is a treasure—a man of great talents and integrity. I met with him here by chance, employed and recommended him, and he has now under his charge all the works of any magnitude in this great and rich district."

Another of our engineer's early stone bridges, which may be mentioned in this place, was erected by him in 1805, over the river Dee at Tongueland in the county of

* His foreman of masons at Bewdley Bridge, and afterwards his assistant in numerous important works.

Kirkcudbright. It is a bold and picturesque bridge, situated in a lovely locality. The river is very deep at high water there, the tide rising 20 feet. As the banks were steep and rocky, the engineer determined to bridge the stream by a single arch of 112 feet span. The rise being

TONGUELAND BRIDGE.

[By R. P. Leitch, after a Drawing by J. S. Smiles.]

considerable, high wingwalls and deep spandrels were requisite; but the weight of the structure was much lightened by the expedient which he adopted of perforating the wings, and building a number of longitudinal walls in the spandrels, instead of filling them with earth or inferior masonry, as had until then been the ordinary practice. The ends of these walls, connected and steadied by the insertion of tee-stones, were built so as to abut against the back of the arch-stones and the cross walls of each abutment. Thus great strength as well as lightness was secured, and a very graceful and at the same time substantial bridge was provided for the accommodation of the district.*

In his letters written about this time, Telford seems to have been very full of employment, which required him to travel about a great deal. "I have become," said he, "a very wandering being, and am scarcely ever two days in one place, unless detained by business, which, however, occupies my time very completely." At another time he says, "I am tossed about like a tennis ball: the other day I was in London, since that I have been in Liverpool, and in a few days I expect to be at Bristol. Such is my life; and to tell you the truth, I think it suits my disposition."

Another work on which Telford was engaged at this time was a project for supplying the town of Liverpool with water conveyed through pipes in the same manner as had long before been adopted in London. He was much struck by the activity and enterprise apparent in Liverpool compared with Bristol. "Liverpool," he said, "has taken firm root in the country by means of the canals: it is young, vigorous, and well situated. Bristol is sinking in commercial importance: its merchants are rich and indolent, and in their projects they are always too late. Besides, the place is badly situated. There will probably arise another port there somewhat nearer the Severn; but Liverpool will nevertheless continue of the first commercial importance, and their water will be turned into wine. We are making rapid progress in this country—I mean from Liver-

* The work is thus described in Robert Chambers's 'Picture of Scotland':—"Opposite Compston there is a magnificent new bridge over the Dee. It consists of a single arch, the span of which is 112 feet; and it is built of vast blocks of freestone brought from the Isle of Arran. The cost of this work was somewhere about 7000*l.* sterling; and it may be mentioned, to the honour of the Stewartry, that this sum was raised by the private contributions of the gentlemen of the district. From Tongueland Hill, in the immediate vicinity of the bridge, there is a view well worthy of a painter's eye, and which is not inferior in beauty and magnificence to any in Scotland."

pool to Bristol, and from Wales to Birmingham. This is an extensive and rich district, abounding in coal, lime, iron, and lead. Agriculture too is improving, and manufactures are advancing at rapid strides towards perfection. Think of such a mass of population, industrious, intelligent, and energetic, in continual exertion! In short, I do not believe that any part of the world, of like dimensions, ever exceeded Great Britain, as it now is, in regard to the production of wealth and the practice of the useful arts." *

Amidst all this progress, which so strikingly characterized the western districts of England, Telford also thought that there was a prospect of coming improvement for Ireland. "There is a board of five members appointed by Parliament, to act as a board of control over all the inland navigations, &c., of Ireland. One of the members is a particular friend of mine, and at this moment a pupil, as it were, anxious for information. This is a noble object: the field is wide, the ground new and capable of vast improvement. To take up and manage the water of a fine island is like a fairy tale, and, if properly conducted, it would render Ireland truly a jewel among the nations." † It does not, however, appear that Telford was ever employed by the board to carry out the grand scheme which thus fired his engineering imagination.

Mixing freely with men of all classes, our engineer seems to have made many new friends and acquaintances about this time. While on his journeys north and south, he frequently took the opportunity of looking in upon the venerable James Watt—"a great and good man," he terms him—at his house at Heathfield, near Birmingham. At London he says he is "often with old Brodie and Black, each the first in his profession, though they walked up together to the great city on foot,‡ more than half a century

* Letter to Mr. Andrew Little, Langholm, dated Salop, 13th July, 1799.

† Letter to Mr. Andrew Little, Langholm, dated Liverpool, 9th September, 1800.

‡ Brodie was originally a blacksmith. He was a man of much

ago—Gloria!" About the same time we find him taking interest in the projects of a deserving person, named Holwell, a coal-master in Staffordshire, and assisting him to take out a patent for boring wooden pipes; "he being a person," says Telford, "little known, and not having capital, interest, or connections, to bring the matter forward."

Telford also kept up his literary friendships and preserved his love for poetical reading. At Shrewsbury, one of his most intimate friends was Dr. Darwin, son of the author of the 'Botanic Garden.' At Liverpool, he made the acquaintance of Dr. Currie, and was favoured with a sight of his manuscript of the 'Life of Burns,' then in course of publication. Curiously enough, Dr. Currie had found among Burns's papers a copy of some verses, addressed to the poet, which Telford recognised as his own, written many years before while working as a mason at Langholm. Their purport was to urge Burns to devote himself to the composition of poems of a serious character, such as the 'Cotter's Saturday Night.' With Telford's permission, several extracts from his Address to Burns were published in 1800 in Currie's Life of the poet. Another of his literary friendships, formed about the same time, was that with Thomas Campbell, then a very young man, whose 'Pleasures of Hope' had just made its appearance. Telford, in one of his letters, says, "I will not leave a stone unturned to try to serve the author of that charming poem." In a subsequent communication * he says, "The author of the 'Pleasures of Hope' has been here for some time. I am quite delighted with him. He is the very spirit of poetry. On Monday I introduced him to

ingenuity and industry, and introduced many improvements in iron work; he invented stoves for chimneys, ships' hearths, &c. He had above a hundred men working in his London shop, besides carrying on an iron work at Coalbrookdale. He afterwards established a woollen manufactory near Peebles.

* Dated London, 14th April, 1802.

the King's librarian, and I imagine some good may result to him from the introduction."

In the midst of his plans of docks, canals, and bridges, he wrote letters to his friends about the peculiarities of Goethe's poems and Kotzebue's plays, Roman antiquities, Buonaparte's campaign in Egypt, and the merits of the last new book. He confessed, however, that his leisure for reading was rapidly diminishing in consequence of the increasing professional demands upon his time; but he bought the 'Encyclopedia Britannica,' which he described as "a perfect treasure, containing everything, and always at hand." He thus rapidly described the manner in which his time was engrossed. "A few days since, I attended a general assembly of the canal proprietors in Shropshire. I have to be at Chester again in a week, upon an arbitration business respecting the rebuilding of the county hall and gaol; but previous to that I must visit Liverpool, and afterwards proceed into Worcestershire. So you see what sort of a life I have of it. It is something like Buonaparte, when in Italy, fighting battles at fifty or a hundred miles distance every other day. However, plenty of employment is what every professional man is seeking after, and my various occupations now require of me great exertions, which they certainly shall have so long as life and health are spared to me." *

Amidst all his engagements, Telford found time to make particular inquiry about many poor families formerly known to him in Eskdale, for some of whom he paid house-rent, while he transmitted the means of supplying others with coals, meal, and necessaries, during the severe winter months,—a practice which he continued to the close of his life.

* Letter to Mr. Andrew Little, Langholm, dated Salop, 30th November, 1799.

CHAPTER VIII.

HIGHLAND ROADS AND BRIDGES.

IN an early chapter of this volume we have given a rapid survey of the state of Scotland about the middle of last century. We found a country without roads, fields lying uncultivated, mines unexplored, and all branches of industry languishing, in the midst of an idle, miserable, and haggard population. Fifty years passed, and the state of the Lowlands had become completely changed. Roads had been made, canals dug, coal-mines opened up, iron-works established; manufactures were extending in all directions; and Scotch agriculture, instead of being the worst, was admitted to be the best in the island.

"I have been perfectly astonished," wrote Romilly from Stirling, in 1793, "at the richness and high cultivation of all the tract of this calumniated country through which I have passed, and which extends quite from Edinburgh to the mountains where I now am. It is true, however, that almost everything which one sees to admire in the way of cultivation is due to modern improvements; and now and then one observes a few acres of brown moss, contrasting admirably with the corn-fields to which they are contiguous, and affording a specimen of the dreariness and desolation which, only half a century ago, overspread a country now highly cultivated, and become a most copious source of human happiness." *

It must, however, be admitted that the industrial progress thus described was confined almost entirely to the Lowlands, and had scarcely penetrated the mountainous

* 'Romilly's Autobiography,' ii. 22.

regions lying towards the north-west. The rugged nature of that part of the country interposed a formidable barrier to improvement, and the district still remained very imperfectly opened up. The only practicable roads were those which had been made by the soldiery after the rebellions of 1715 and '45, through counties which before had been inaccessible except by dangerous footpaths across high and rugged mountains. An old epigram in vogue at the end of last century ran thus:—

> "Had you seen these roads before they were made,
> You'd lift up your hands and bless General Wade!"

Being constructed by soldiers for military purposes, they were first known as "military roads." One was formed along the Great Glen of Scotland, in the line of the present Caledonian Canal, connected with the Lowlands by the road through Glencoe by Tyndrum down the western banks of Loch Lomond; another, more northerly, connected Fort Augustus with Dunkeld by Blair Athol; while a third, still further to the north and east, connected Fort George with Cupar-in-Angus by Badenoch and Braemar.

The military roads were about eight hundred miles in extent, and maintained at the public expense. But they were laid out for purposes of military occupation rather than for the convenience of the districts which they traversed. Hence they were comparatively little used, and the Highlanders, in passing from one place to another, for the most part continued to travel by the old cattle tracks along the mountains. But the population were as yet so poor and so spiritless, and industry was in so backward a state all over the Highlands, that the want of more convenient communications was scarcely felt.

Though there was plenty of good timber in certain districts, the bark was the only part that could be sent to market, on the backs of ponies, while the timber itself was left to rot upon the ground. Agriculture was in a surprisingly backward state. In the remoter districts only a little oats or barley was grown, the chief part of which was required for the sustenance of the cattle during winter. The Rev. Mr. Macdougall, minister of the parishes of Lochgoilhead and Kilmorich, in Argyleshire, described the people of that part of the country, about the year 1760, as miserable beyond description. He says, "Indolence was almost the only comfort they enjoyed. There was scarcely any variety of wretchedness with which they were not obliged to struggle, or rather to which they were not obliged to submit. They often felt what it was to want food. . . . To such an extremity were they frequently reduced, that they were obliged to bleed their cattle, in order to subsist some time on the blood (boiled); and even the inhabitants of the glens and valleys repaired in crowds to the shore, at the distance of three or four miles, to pick up the scanty provision which the shell-fish afforded them." *

The plough had not yet penetrated into the Highlands; an instrument called the cas-chrom †—literally the "crooked-

---

* 'Statistical Account of Scotland,' iii. 185.

† The *cas-chrom* was a rude combination of a lever for the removal of rocks, a spade to cut the earth, and a foot-plough to turn it. We annex an illustration of this curious and now obsolete instrument. It weighed about eighteen pounds. In working it, the upper part of the handle, to which the left hand was applied, reached the workman's shoulder, and being slightly elevated, the point, shod with iron, was pushed into the ground horizontally; the soil being turned over by inclining the handle to the furrow side, at the same time making the heel act as a fulcrum to raise the point of the instrument. In turning up unbroken ground, it was first employed with the heel uppermost, with pushing strokes to cut the breadth of the sward to be turned over; after which, it was used horizontally as above described. We are indebted to a Parliamentary Blue Book for the following representation of this interesting relic of ancient agri-

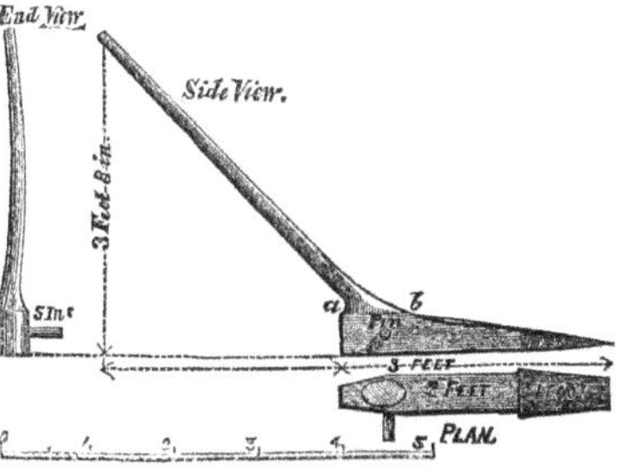

THE CAS-CHROM.

foot"—the use of which had been forgotten for hundreds of years in every other country in Europe, was almost the only tool employed in tillage in those parts of the Highlands which were separated by almost impassable mountains from the rest of the United Kingdom.

The native population were by necessity peaceful. Old feuds were restrained by the strong arm of the law, if indeed the spirit of the clans had not been completely broken by the severe repressive measures which followed the rebellion of Forty-five. But the people had not yet learnt to bend their backs, like the Sassenach, to the stubborn soil, and they sat gloomily by their turf-fires at home, or wandered away to settle in other lands beyond the seas. It even began to be feared that the country would soon be entirely depopulated; and it became a matter of national concern to devise methods of opening up the district so as to develope its industry and afford improved means of sustenance for its population. The poverty of the inhabitants rendered the attempt to construct roads—even had they desired them—beyond their scanty means; but the ministry of the day entertained the opinion that, by contributing a certain proportion of the necessary expense, the proprietors of Highland estates might be induced to advance the remainder; and on this principle the construction of the new roads in those districts was undertaken.

The country lying to the west of the Great Glen was absolutely without a road of any kind. The only district through which travellers passed was that penetrated by the great Highland road by Badenoch, between Perth and Inverness; and for a considerable time after the suppression of the rebellion of 1745, it was infested by gangs of desperate robbers. So unsafe was the route across the Grampians, that persons who had occasion to travel it

culture. It is given in the appendix to the 'Ninth Report of the Commissioners for Highland Roads and Bridges,' ordered by the House of Commons to be printed, 19th April, 1821.

usually made their wills before setting out. Garrons, or little Highland ponies, were then used by the gentry as well as the peasantry. Inns were few and bad; and even when postchaises were introduced at Inverness, the expense of hiring one was thought of for weeks, perhaps months, and arrangements were usually made for sharing it among as many individuals as it would contain. If the harness and springs of the vehicle held together, travellers thought themselves fortunate in reaching Edinburgh, jaded and weary, but safe in purse and limb, on the eighth day after leaving Inverness.* Very few persons then travelled into the Highlands on foot, though Bewick, the father of wood-engraving, made such a journey round Loch Lomond in 1775. He relates that his appearance excited the greatest interest at the Highland huts in which he lodged, the women curiously examining him from head to foot, having never seen an Englishman before. The strange part of his story is, that he set out upon his journey from Cherryburn, near Newcastle, with only three guineas sewed in his waistband, and when he reached home he had still a few shillings left in his pocket!

In 1802, Mr. Telford was called upon by the Government to make a survey of Scotland, and report as to the measures which were necessary for the improvement of the roads and bridges of that part of the kingdom, and also on the means of promoting the fisheries on the east and west coasts, with the object of better opening up the country and preventing further extensive emigration. Previous to this time he had been employed by the British Fisheries Society—of which his friend Sir William Pulteney was Governor—to inspect the harbours at their several stations, and to devise a plan for the establishment of a fishery on the coast of Caithness. He accordingly made an extensive tour of Scotland, examining, among

* Anderson's 'Guide to the Highlands and Islands of Scotland,' 3rd ed. p. 48.

other harbours, that of Annan; from which he proceeded northward by Aberdeen to Wick and Thurso, returning to Shrewsbury by Edinburgh and Dumfries.* He accumulated a large mass of data for his report, which was sent in to the Fishery Society, with charts and plans, in the course of the following year.

In July, 1802, he was requested by the Lords of the Treasury, most probably in consequence of the preceding report, to make a further survey of the interior of the Highlands, the result of which he communicated in his report presented to Parliament in the following year. Although full of important local business, "kept running," as he says, "from town to country, and from country to town, never when awake, and perhaps not always when asleep, have my Scotch surveys been absent from my mind." He had worked very hard at his report, and hoped that it might be productive of some good.

The report was duly presented, printed,† and approved; and it formed the starting-point of a system of legislation with reference to the Highlands which extended over many years, and had the effect of completely opening up that romantic but rugged district of country, and extending to its inhabitants the advantages of improved intercourse with the other parts of the kingdom. Mr. Telford pointed out that the military roads were altogether inadequate to the requirements of the population, and that the use of them was in many places very much circumscribed by the want of bridges over some of the principal rivers. For instance, the route from Edinburgh to Inver-

* He was accompanied on this tour by Colonel Dirom, with whom he returned to his house at Mount Annan, in Dumfries. Telford says of him: "The Colonel seems to have roused the county of Dumfries from the lethargy in which it has slumbered for centuries. The map of the county, the mineralogical survey, the new roads, the opening of lime works, the competition of ploughing, the improving harbours, the building of bridges, are works which bespeak the exertions of no common man."—Letter to Mr. Andrew Little, dated Shrewsbury, 30th November, 1801.

† Ordered to be printed 5th of April, 1803.

ness, through the Central Highlands, was seriously interrupted at Dunkeld, where the Tay is broad and deep, and not always easy to be crossed by means of a boat. The route to the same place by the east coast was in like manner broken at Fochabers, where the rapid Spey could only be crossed by a dangerous ferry.

The difficulties encountered by gentlemen of the Bar, in travelling the north circuit about this time, are well described by Lord Cockburn in his 'Memorials.' "Those who are born to modern travelling," he says, "can scarcely be made to understand how the previous age got on. The state of the roads may be judged of from two or three facts. There was no bridge over the Tay at Dunkeld, or over the Spey at Fochabers, or over the Findhorn at Forres. Nothing but wretched pierless ferries, let to poor cottars, who rowed, or hauled, or pushed a crazy boat across, or more commonly got their wives to do it. There was no mail-coach north of Aberdeen till, I think, after the battle of Waterloo. What it must have been a few years before my time may be judged of from Bozzy's 'Letter to Lord Braxfield,' published in 1780. He thinks that, besides a carriage and his own carriage-horses, every judge ought to have his sumpter-horse, and ought not to travel faster than the waggon which carried the baggage of the circuit. I understood from Hope that, after 1784, when he came to the Bar, he and Braxfield *rode* a whole north circuit; and that, from the Findhorn being in a flood, they were obliged to go up its banks for about twenty-eight miles to the bridge of Dulsie before they could cross. I myself rode circuits when I was Advocate-Depute between 1807 and 1810. The fashion of every Depute carrying his own shell on his back, in the form of his own carriage, is a piece of very modern antiquity." *

North of Inverness, matters were, if possible, still worse. There was no bridge over the Beauly or the Conan. The

* 'Memorials of his Time,' by Henry Cockburn, pp. 341-3.

drovers coming south swam the rivers with their cattle. There being no roads, there was little use for carts. In the whole county of Caithness, there was scarcely a farmer who owned a wheel-cart. Burdens were conveyed usually on the backs of ponies, but quite as often on the backs of women.* The interior of the county of Sutherland being almost inaccessible, the only track lay along the shore, among rocks and sand, and was covered by the sea at every tide. "The people lay scattered in inaccessible straths and spots among the mountains, where they lived in family with their pigs and kyloes (cattle), in turf cabins of the most miserable description; they spoke only Gaelic, and spent the whole of their time in indolence and sloth. Thus they had gone on from father to son, with little change, except what the introduction of illicit distillation had wrought, and making little or no export from the country beyond the few lean kyloes, which paid the rent and produced wherewithal to pay for the oatmeal imported." †

Telford's first recommendation was, that a bridge should be thrown across the Tay at Dunkeld, to connect the improved lines of road proposed to be made on each side of the river. He regarded this measure as of the first importance to the Central Highlands; and as the Duke of Athol was willing to pay one-half of the cost of the erection, if the Government would defray the other—the bridge to be free of toll after a certain period—it appeared to the engineer that this was a reasonable and just mode of providing for the contingency. In the next place, he recommended a bridge over the Spey, which drained a great extent of mountainous country, and, being liable to sudden inundations, was very dangerous to cross. Yet this ferry formed the only link of communication between

* 'Memoirs of the Life and Writings of Sir John Sinclair, Bart.,' vol. i., p. 339.

† Extract of a letter from a gentleman residing in Sunderland, quoted in 'Life of Telford,' p. 465.

the whole of the northern counties. The site pointed out for the proposed bridge was adjacent to the town of Fochabers, and here also the Duke of Gordon and other county gentlemen were willing to provide one-half of the means for its erection.

Mr. Telford further described in detail the roads necessary to be constructed in the north and west Highlands, with the object of opening up the western parts of the counties of Inverness and Ross, and affording a ready communication from the Clyde to the fishing lochs in the neighbourhood of the Isle of Skye. As to the means of executing these improvements, he suggested that Government would be justified in dealing with the Highland roads and bridges as exceptional and extraordinary works, and extending the public aid towards carrying them into effect, as, but for such assistance, the country must remain, perhaps for ages to come, imperfectly opened up. His report further embraced certain improvements in the harbours of Aberdeen and Wick, and a description of the country through which the proposed line of the Caledonian Canal would necessarily pass—a canal which had long been the subject of inquiry, but had not as yet emerged from a state of mere speculation.

The new roads, bridges, and other improvements suggested by the engineer, excited much interest in the north. The Highland Society voted him their thanks by acclamation; the counties of Inverness and Ross followed; and he had letters of thanks and congratulation from many of the Highland chiefs. "If they will persevere," says he, "with anything like their present zeal, they will have the satisfaction of greatly improving a country that has been too long neglected. Things are greatly changed now in the Highlands. Even were the chiefs to quarrel, de'il a Highlandman would stir for them. The lairds have transferred their affections from their people to flocks of sheep, and the people have lost their veneration for the lairds. It seems to be the natural progress of society; but it is not an altogether satisfactory change. There were some fine features in the former patriarchal state of society; but now clanship is gone, and chiefs and people are hastening into the opposite extreme. This seems to me to be quite wrong."*

In the same year, Telford was elected a member of the Royal Society of Edinburgh, on which occasion he was proposed and supported by three professors; so that the former Edinburgh mason was rising in the world and receiving due honour in his own country. The effect of his report was such, that in the session of 1803 a Parliamentary Commission was appointed, under whose direction a series of practical improvements was commenced, which issued in the construction of not less than 920 additional miles of roads and bridges throughout the Highlands, one-half of the cost of which was defrayed by the Government and the other half by local assessment. But in addition to these main lines of communication, numberless county roads were formed by statute labour, under local road Acts and by other means; the land-owners of Sutherland alone constructing nearly 300 miles of district roads at their own cost.

By the end of the session of 1803, Telford received his instructions from Mr. Vansittart as to the working survey he was forthwith required to enter upon, with a view to commencing practical operations; and he again proceeded to the Highlands to lay out the roads and plan the bridges which were most urgently needed. The district of the Solway was, at his representation, included, with the object of improving the road from Carlisle to Portpatrick—the nearest point at which Great Britain meets the Irish coast, and where the sea passage forms only a sort of wide ferry.

It would occupy too much space, and indeed it is altogether unnecessary, to describe in detail the operations of the Commission and of their engineer in opening up the

* Letter to Mr. Andrew Little, Langholm, dated Salop, 18th February, 1803.

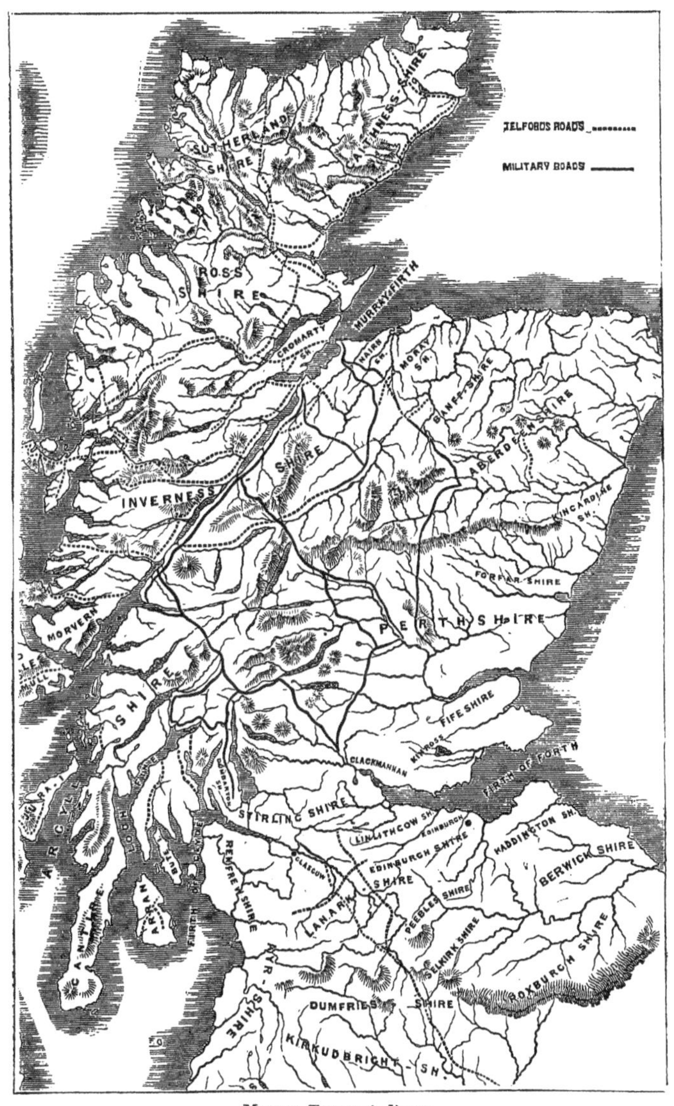

MAP OF TELFORD'S ROADS.

communications of the Highlands. Suffice it to say, that one of the first things taken in hand was the connection of the existing lines of road by means of bridges at the more important points; such as at Dunkeld over the Tay, and near Dingwall over the Conan and Orrin. That of Dunkeld was the most important, as being situated at the

DUNKELD BRIDGE. [By Percival Skelton.]

entrance to the Central Highlands; and at the second meeting of the Commissioners Mr. Telford submitted his plan and estimates of the proposed bridge. In consequence of some difference with the Duke of Athol as to his share of the expense—which proved to be greater than he had estimated—some delay occurred in beginning the work; but at length it was fairly started, and, after being about three years in hand, the structure was finished and opened for traffic in 1809.

The bridge is a handsome one of five river and two land arches. The span of the centre arch is 90 feet. of the two adjoining it 84 feet, and of the two side arches 74 feet; affording a clear waterway of 446 feet. The total breadth of the roadway and footpaths is 28 feet 6 inches. The cost of the structure was about 14,000*l.*, one-half of which was defrayed by the Duke of Athol. Dunkeld bridge now forms a fine feature in a landscape not often surpassed, and which presents within a comparatively small compass a great variety of character and beauty.

The communication by road north of Inverness was also perfected by the construction of a bridge of five arches over the Beauly, and another of the same number over the Conan, the central arch being 65 feet span; and the formerly wretched bit of road between these points having been put in good repair, the town of Dingwall was thenceforward rendered easily approachable from the south. At the same time, a beginning was made with the construction of new roads through the districts most in need of them. The first contracted for, was the Loch-na-Gaul road, from Fort William to Arasaig, on the western coast, nearly opposite the island of Egg. Another was begun from Loch Oich, on the line of the Caledonian Canal, across the middle of the Highlands, through Glengarry, to Loch Hourn on the western sea. Other roads were opened north and south; through Morvern to Loch Moidart; through Glen Morrison and Glen Sheil, and through the entire Isle of Skye; from Dingwall, eastward, to Lochcarron and Loch Torridon, quite through the county of Ross; and from Dingwall, northward, through the county of Sutherland as far as Tongue on the Pentland Frith; while another line, striking off at the head of the Dornoch Frith, proceeded along the coast in a north-easterly direction to Wick and Thurso, in the immediate neighbourhood of John o' Groats.

There were numerous other subordinate lines of road which it is unnecessary to specify in detail; but some idea may be formed of their extent, as well as of the rugged character of the country through which they were carried, when we state that they involved the construction of no fewer than twelve hundred bridges. Several important bridges were also erected at other points to connect existing roads, such as those at Ballater and Potarch over the Dee; at Alford over the Don: and at Craig-Ellachie over the Spey.

The last named bridge is a remarkably elegant structure, thrown over the Spey at a point where the river, rushing obliquely against the lofty rock of Craig-Ellachie,* has formed for itself a deep channel not exceeding fifty yards in breadth. Only a few years before, there had not been any provision for crossing this river at its lower parts except the very dangerous ferry at Fochabers. The Duke of Gordon had, however, erected a suspension bridge at that town, and the inconvenience was in a great measure removed. Its utility was so generally felt, that the demand arose for a second bridge across the river; for there was not another by which it could be crossed for a distance of nearly fifty miles up Strath Spey.

It was a difficult stream to span by a bridge at any place, in consequence of the violence with which the floods descended at particular seasons. Sometimes, even in summer, when not a drop of rain had fallen, the flood would come down the Strath in great fury, sweeping everything before it; this remarkable phenomenon being accounted for by the prevalence of a strong south-westerly wind, which blew the loch waters from their beds into the Strath, and thus suddenly filled the valley of the Spey.† The same

* The names of Celtic places are highly descriptive. Thus *Craig-Ellachie* literally means, the rock of separation; *Badenoch*, bushy or woody; *Cairngorm*, the blue cairn; *Lochinet*, the lake of nests; *Bal-knockan*, the town of knolls; *Dalna-sealg*, the hunting dale; *Alt'n dater*, the burn of the horn-blower; and so on.

† Sir Thomas Dick Lauder has vividly described the destructive character of the Spey-side inundations in his capital book on the 'Morayshire Floods.'

phenomenon, similarly caused, is also frequently observed in the neighbouring river, the Findhorn, cooped up in its deep rocky bed, where the water sometimes comes down in a wave six feet high, like a liquid wall, sweeping everything before it.

To meet such a contingency, it was deemed necessary to provide abundant waterway, and to build a bridge offering

CRAIG-ELLACHIE BRIDGE. [By Percival Skelton.]

as little resistance as possible to the passage of the Highland floods. Telford accordingly designed for the passage of the river at Craig-Ellachie a light cast-iron arch of 150 feet span, with a rise of 20 feet, the arch being composed of four ribs, each consisting of two concentric arcs forming panels, which are filled in with diagonal bars.

The roadway is 15 feet wide, and is formed of another arc of greater radius, attached to which is the iron railing; the spandrels being filled by diagonal ties, forming trellis-work. Mr. Robert Stephenson took objection to the two dissimilar arches, as liable to subject the structure, from variations of temperature, to very unequal strains. Nevertheless this bridge, as well as many others constructed by Mr. Telford after a similar plan, has stood perfectly well, and to this day remains a very serviceable structure.

Its appearance is highly picturesque. The scattered pines and beech trees on the side of the impending mountain, the meadows along the valley of the Spey, and the western approach road to the bridge cut deeply into the face of the rock, combine, with the slender appearance of the iron arch, in rendering this spot one of the most remarkable in Scotland.*

An iron bridge of a similar span to that at Craig-Ellachie had previously been constructed across the head of the Dornoch Frith at Bonar, near the point where the waters of the Shin join the sea. The very severe trial which this structure sustained from the tremendous blow of an irregular mass of fir-tree logs, consolidated by ice, as well as, shortly after, from the blow of a schooner which drifted against it on the opposite side, and had her two masts knocked off by the collision, gave him every confidence in the strength of this form of construction, and he accordingly repeated it in several of his subsequent bridges, though none of them are comparable in beauty with that of Craig-Ellachie.

Thus, in the course of eighteen years, 920 miles of capital roads, connected together by no fewer than 1200 bridges, were added to the road communications of the Highlands, at an expense defrayed partly by the localities immediately benefited, and partly by the nation. The

---

* 'Report of the Commissioners on Highland Roads and Bridges. Appendix to 'Life of Telford,' p. 400.

effects of these twenty years' operations were such as follow the making of roads everywhere—development of industry and increase of civilization. In no districts were the benefits derived from them more marked than in the remote northern counties of Sutherland and Caithness. The first stage-coaches that ran northward from Perth to Inverness were tried in 1806, and became regularly established in 1811; and by the year 1820 no fewer than forty arrived at the latter town in the course of every week, and the same number departed from it. Others were established in various directions through the highlands, which were rendered as accessible as any English county.

Agriculture máde rapid progress. The use of carts became practicable, and manure was no longer carried to the field on women's backs. Sloth and idleness gradually disappeared before the energy, activity, and industry which were called into life by the improved communications. Better built cottages took the place of the old mud biggins with holes in their roofs to let out the smoke. The pigs and cattle were treated to a separate table. The dunghill was turned to the outside of the house. Tartan tatters gave place to the produce of Manchester and Glasgow looms; and very soon few young persons were to be found who could not both read and write English.

But not less remarkable were the effects of the road-making upon the industrial habits of the people. Before Telford went into the Highlands, they did not know how to work, having never been accustomed to labour continuously and systematically. Let our engineer himself describe the moral influences of his Highland contracts:—
"In these works," says he, "and in the Caledonian Canal, about three thousand two hundred men have been annually employed. At first, they could scarcely work at all: they were totally unacquainted with labour; they could not use the tools. They have since become excellent labourers, and of the above number we consider about one-fourth left us annually, taught to work. These undertakings may,

indeed, be regarded in the light of a working academy, from which eight hundred men have annually gone forth improved workmen. They have either returned to their native districts with the advantage of having used the most perfect sort of tools and utensils (which alone cannot be estimated at less than ten per cent. on any sort of labour), or they have been usefully distributed through the other parts of the country. Since these roads were made accessible, wheelwrights and cartwrights have been established, the plough has been introduced, and improved tools and utensils are generally used. The plough was not previously employed; in the interior and mountainous parts they used crooked sticks, with iron on them, drawn or pushed along. The moral habits of the great masses of the working classes are changed; they see that they may depend on their own exertions for support: this goes on silently, and is scarcely perceived until apparent by the results. I consider these improvements among the greatest blessings ever conferred on any country. About two hundred thousand pounds has been granted in fifteen years. It has been the means of advancing the country at least a century."

The progress made in the Lowland districts of Scotland since the same period has been no less remarkable. If the state of the country, as we have above described it from authentic documents, be compared with what it is now, it will be found that there are few countries which have accomplished so much within so short a period. It is usual to cite the United States as furnishing the most extraordinary instance of social progress in modern times. But America has had the advantage of importing its civilization for the most part ready made, whereas that of Scotland has been entirely her own creation. By nature America is rich, and of boundless extent; whereas Scotland is by nature poor, the greater part of her limited area consisting of sterile heath and mountain. Little more than a century ago, Scotland was considerably in the rear of Ireland. It was a country almost without agriculture, without mines, without fisheries, without shipping, without money, without roads. The people were ill-fed, half barbarous, and habitually indolent. The colliers and salters were veritable slaves, and were subject to be sold together with the estates to which they belonged.

What do we find now? Prædial slavery completely abolished; heritable jurisdictions at an end; the face of the country entirely changed; its agriculture acknowledged to be the first in the world; its mines and fisheries productive in the highest degree; its banking a model of efficiency and public usefulness; its roads equal to the best roads in England or in Europe. The people are active and energetic, alike in education, in trade, in manufactures, in construction, in invention. Watt's invention of the steam-engine, and Symington's invention of the steam-boat, proved a source of wealth and power, not only to their own country, but to the world at large; while Telford, by his roads, bound England and Scotland, before separated, firmly into one, and rendered the union a source of wealth and strength to both.

At the same time, active and powerful minds were occupied in extending the domain of knowledge,—Adam Smith in Political Economy, Reid and Dugald Stewart in Moral Philosophy, and Black and Robison in Physical Science. And thus Scotland, instead of being one of the idlest and most backward countries in Europe, has, within the compass of little more than a lifetime, issued in one of the most active, contented, and prosperous,—exercising an amount of influence upon the literature, science, political economy, and industry of modern times, out of all proportion to the natural resources of its soil or the amount of its population.

If we look for the causes of this extraordinary social progress, we shall probably find the principal to consist in the fact that Scotland, though originally poor as a country, was rich in parish schools, founded under the provisions of an Act passed by the Scottish Parliament in the year 1696. It was there ordained, "that there be a school settled and established, and a schoolmaster appointed, in every parish not already provided, by advice of the heritors and minister of the parish." Common day-schools were accordingly provided and maintained throughout the country for the education of children of all ranks and conditions. The consequence was, that in the course of a few generations, these schools, working steadily upon the minds of the young, all of whom passed under the hands of the teachers, educated the population into a state of intelligence and aptitude greatly in advance of their material well-being; and it is in this circumstance, we apprehend, that the explanation is to be found of the rapid start forward which the whole country took, dating more particularly from the year 1745. Agriculture was naturally the first branch of industry to exhibit signs of decided improvement; to be speedily followed by like advances in trade, commerce, and manufactures. Indeed, from that time the country never looked back, but her progress went on at a constantly accelerated rate, issuing in results as marvellous as they have probably been unprecedented.

CHAPTER IX.

TELFORD'S SCOTCH HARBOURS.

No sooner were the Highland roads and bridges in full progress, than attention was directed to the improvement of the harbours round the coast. Very little had as yet been done for them beyond what nature had effected. Happily, there was a public fund at disposal—the accumulation of rents and profits derived from the estates forfeited at the rebellion of 1745—which was available for the purpose. The suppression of the rebellion did good in many ways. It broke the feudal spirit, which lingered in the Highlands long after it had ceased in every other part of Britain; it led to the effectual opening up of the country by a system of good roads; and now the accumulated rents of the defeated Jacobite chiefs were about to be applied to the improvement of the Highland harbours for the benefit of the general population.

The harbour of Wick was one of the first to which Mr. Telford's attention was directed. Mr. Rennie had reported on the subject of its improvement as early as the year 1793, but his plans were not adopted because their execution was beyond the means of the locality at that time. The place had now, however, become of considerable importance. It was largely frequented by Dutch fishermen during the herring season; and it was hoped that, if they could be induced to form a settlement at the place, their example might exercise a beneficial influence upon the population.

Mr. Telford reported that, by the expenditure of about 5890*l.*, a capacious and well-protected tidal basin might be formed, capable of containing about two hundred herring-busses. The Commission adopted his plan, and voted the requisite funds for carrying out the works, which were begun in 1808. The new station was named Pulteney Town, in compliment to Sir William Pulteney, the Governor of the Fishery Society; and the harbour was built at a cost of about 12,000*l.*, of which 8500*l.* was granted from the Forfeited Estates Fund. A handsome stone bridge, erected over the River Wick in 1805, after the design of our engineer, connects these improvements with the older town: it is formed of three arches, having a clear waterway of 156 feet.

The money was well expended, as the result proved; and Wick is now, we believe, the greatest fishing station in the world. The place has increased from a little poverty-stricken village to a large and thriving town, which swarms during the fishing season with lowland Scotchmen, fair Northmen, broad-built Dutchmen, and kilted Highlanders. The bay is at that time frequented by upwards of a thousand fishing-boats, and the take of herrings in some years amounts to more than a hundred thousand barrels. The harbour has of late years been considerably improved to meet the growing requirements of the herring trade, the principal additions having been carried out, in 1823, by Mr. Bremner,* a native engineer of great ability.

* Hugh Millar, in his 'Cruise of the Betsy,' attributes the invention of columnar pier-work to Mr. Bremner, whom he terms "the Brindley of Scotland." He has acquired great fame for his skill in raising sunken ships, having warped the *Great Britain* steamer off the shores of Dundrum Bay. But we believe Mr. Telford had adopted the practice of columnar pier-work before Mr. Bremner, in forming the little harbour of Folkestone in 1808, where the work is still to be seen quite perfect. The most solid mode of laying stone on land is in flat courses; but in open pier work the reverse process is adopted. The blocks are laid on end in columns, like upright beams jammed together. Thus laid, the wave which dashes against them is broken, and spends itself on the interstices; whereas, if it struck the broad solid blocks, the tendency would be to lift them from their beds and set the work afloat; and in a furious storm such blocks would be driven about almost like pebbles. The rebound from flat surfaces is also very heavy, and produces violent commotion; where-

Improvements of a similar kind were carried out by the Fishery Board at other parts of the coast, and many snug and convenient harbours were provided at the principal fishing stations in the Highlands and Western Islands. Where the local proprietors were themselves found expending money in carrying out piers and harbours, the Board assisted them with grants to enable the works to be constructed in the most substantial manner and after the most approved plans. Thus, along that part of the bold northern coast of the mainland of Scotland which projects into the German Ocean, many old harbours were improved or new ones constructed—as at Peterhead, Frazerburgh, Banff, Cullen, Burgh Head, and Nairn. At Fortrose, in the Murray Frith; at Dingwall, in the Cromarty Frith; at Portmaholmac, within Tarbet Ness, the remarkable headland of the Frith of Dornoch; at Kirkwall, the principal town and place of resort in the Orkney Islands, so well known from Sir Walter Scott's description of it in the 'Pirate;' at Tobermory, in the island of Mull;

as these broken, upright, columnar-looking piers seem to absorb the fury of the sea, and render its wildest waves comparatively innocuous.

FOLKESTONE HARBOUR. [By Percival Skelton.]

and at other points of the coast, piers were erected and other improvements carried out to suit the convenience of the growing traffic and trade of the country.

The principal works were those connected with the harbours situated upon the line of coast extending from the harbour of Peterhead, in the county of Aberdeen, round to the head of the Murray Frith. The shores there are exposed to the full force of the seas rolling in from the Northern Ocean; and safe harbours were especially needed for the protection of the shipping passing from north to south. Wrecks had become increasingly frequent, and harbours of refuge were loudly called for. At one part of the coast, as many as thirty wrecks had occurred within a very short time, chiefly for want of shelter.

The situation of Peterhead peculiarly well adapted it for a haven of refuge, and the improvement of the port was early regarded as a matter of national importance. Not far from it, on the south, are the famous Bullars or Boilers of Buchan—bold rugged rocks, some 200 feet high, against which the sea beats with great fury, boiling and churning in the deep caves and recesses with which they are perforated. Peterhead stands on the most easterly part of the mainland of Scotland, occupying the north-east side of the bay, and being connected with the country on the north-west by an isthmus only 800 yards broad. In Cromwell's time, the port possessed only twenty tons of boat tonnage, and its only harbour was a small basin dug out of the rock. Even down to the close of the sixteenth century the place was but an insignificant fishing village. It is now a town bustling with trade, having long been the principal seat of the whale fishery, 1500 men of the port being engaged in that pursuit alone; and it sends out ships of its own building to all parts of the world, its handsome and commodious harbours being accessible at all winds to vessels of almost the largest burden.

It may be mentioned that about sixty years since, the port was formed by the island called Keith Island, situated

PETERHEAD.

By R. P. LEITCH

Page 212.

a small distance eastward from the shore, between which and the mainland an arm of the sea formerly passed. A causeway had, however, been formed across this channel, thus dividing it into two small bays; after which the southern one had been converted into a harbour by means of two rude piers erected along either side of it. The north inlet remained without any pier, and being very inconvenient and exposed to the north-easterly winds, it was little used.

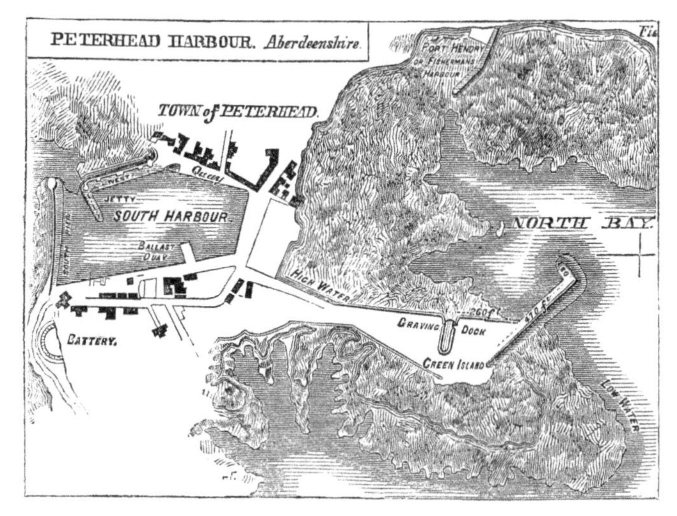

The first works carried out at Peterhead were of a comparatively limited character, the old piers of the south harbour having been built by Smeaton; but improvements proceeded apace with the enterprise and wealth of the inhabitants. Mr. Rennie, and after him Mr. Telford, fully reported as to the capabilities of the port and the best means of improving it. Mr. Rennie recommended the deepening of the south harbour and the extension of the jetty of the west pier, at the same time cutting off all projections of rock from Keith Island on the eastward, so as to render the access more easy. The harbour, when thus finished, would, he estimated, give about 17 feet depth at high water of spring tides. He also proposed to open a communication across the causeway between the north and south harbours, and form a wet dock between them, 580 feet long and 225 feet wide, the water being kept in by gates at each end. He further proposed to provide an entirely new harbour, by constructing two extensive piers for the effectual protection of the northern part of the channel, running out one from a rock north of the Green Island, about 680 feet long, and another from the Roan Head, 450 feet long, leaving an opening between them of 70 yards. This comprehensive plan unhappily could not be carried out at the time for want of funds; but it may be said to have formed the groundwork of all that has been subsequently done for the improvement of the port of Peterhead.

It was resolved, in the first place, to commence operations by improving the south harbour, and protecting it more effectually from south-easterly winds. The bottom of the harbour was accordingly deepened by cutting out 30,000 cubic yards of rocky ground; and part of Mr. Rennie's design was carried out by extending the jetty of the west pier, though only for a distance of twenty yards. These works were executed under Mr. Telford's directions; they were completed by the end of the year 1811, and proved to be of great public convenience.

The trade of the town, however, so much increased, and the port was found of such importance as a place of refuge for vessels frequenting the north seas, that in 1816 it was determined to proceed with the formation of a harbour on the northern part of the old channel; and the inhabitants having agreed among themselves to contribute to the extent of 10,000*l.* towards carrying out the necessary works, they applied for the grant of a like sum from the Forfeited Estates Fund, which was eventually voted for the purpose. The plan adopted was on a more limited scale than that proposed by Mr. Rennie; but in the same direction and contrived with the same object,—so that, when completed, vessels of the largest burden employed in the Greenland fishery might be able to enter one or other of the two harbours and find safe shelter, from whatever quarter the wind might blow.

The works were vigorously proceeded with, and had made considerable progress, when, in October, 1819, a violent hurricane from the north-east, which raged along the coast for several days, and inflicted heavy damage on many of the northern harbours, destroyed a large part of the unfinished masonry and hurled the heaviest blocks into the sea, tossing them about as if they had been pebbles. The finished work had, however, stood well, and the foundations of the piers under low water were ascertained to have remained comparatively uninjured. There was no help for it but to repair the damaged work, though it involved a heavy additional cost, one-half of which was borne by the Forfeited Estates Fund and the remainder by the inhabitants. Increased strength was also given to the more exposed parts of the pierwork, and the slope at the sea side of the breakwater was considerably extended.* Those alterations in the design were carried out, together with a spacious graving-dock, as shown in the preceding plan, and they proved completely successful, enabling Peterhead to offer an amount of accommodation for shipping of a more effectual kind than was at that time to be met with along the whole eastern coast of Scotland.

The old harbour of Frazerburgh, situated on a projecting point of the coast at the foot of Mount Kennaird, about twenty miles north of Peterhead, had become so ruinous that vessels lying within it received almost as little shelter as if they had been exposed in the open sea. Mr. Rennie had prepared a plan for its improvement by running out a substantial north-eastern pier; and this was eventually

* 'Memorials from Peterhead and Banff, concerning Damage occasioned by a Storm.' Ordered by the House of Commons to be printed, 5th July, 1820. [242.]

carried out by Mr. Telford in a modified form, proving of substantial service to the trade of the port. Since then a large and commodious new harbour has been formed at the place, partly at the public expense and partly at that of the inhabitants, rendering Frazerburgh a safe retreat for vessels of war as well as merchantmen.

Among the other important harbour works on the north-east coast carried out by Mr. Telford under the Commissioners appointed to administer the funds of the Forfeited Estates, were those at Banff, the execution of which extended over many years; but, though costly, they did not prove of anything like the same convenience as those executed at Peterhead. The old harbour at the end of the

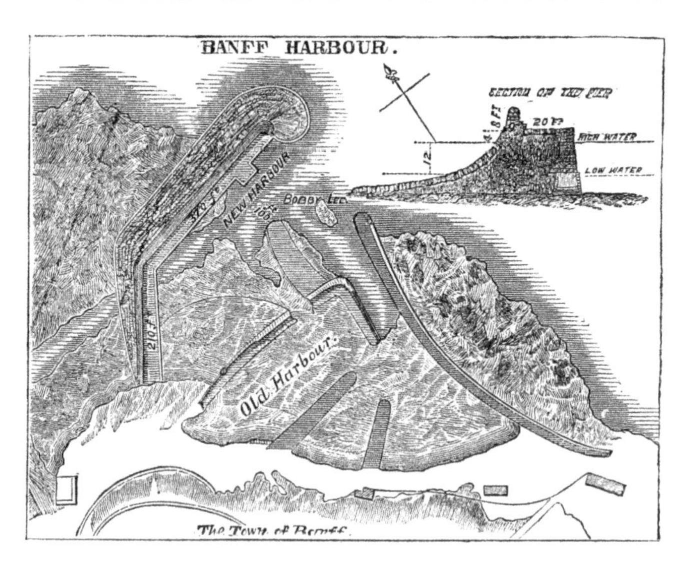

BANFF HARBOUR.

ridge running north and south, on which what is called the "sea town" of Banff is situated, was completed in 1775, when the place was already considered of some importance as a fishing station.

This harbour occupies the triangular space at the north-

BANFF.

By R. P. LEITCH.

Page 216.

eastern extremity of the projecting point of land, at the opposite side of which, fronting the north-west, is the little town and harbour of Macduff. In 1816, Mr. Telford furnished the plan of a new pier and breakwater, covering the old entrance, which presented an opening to the N.N.E., with a basin occupying the intermediate space. The inhabitants agreed to defray one half of the necessary cost, and the Commissioners the other; and the plans having been approved, the works were commenced in 1818. They were in full progress when, unhappily, the same hurricane which in 1819 did so much injury to the works at Peterhead, also fell upon those at Banff, and carried away a large part of the unfinished pier. This accident had the effect of interrupting the work, as well as increasing its cost; but the whole was successfully completed by the year 1822. Although the new harbour did not prove very safe, and exhibited a tendency to become silted up with sand, it proved of use in many respects, more particularly in preventing all swell and agitation in the old harbour, which was thereby rendered the safest artificial haven in the Murray Firth.

It is unnecessary to specify the alterations and improvements of a similar character, adapted to the respective localities, which were carried out by our engineer at Burgh Head, Nairn, Kirkwall, Tarbet, Tobermory, Portmaholmac, Dingwall (with its canal two thousand yards long, connecting the town in a complete manner with the Frith of Cromarty), Cullen, Fortrose, Ballintraed, Portree, Jura, Gourdon, Invergordon, and other places. Down to the year 1823, the Commissioners had expended 108,530*l.* on the improvements of these several ports, in aid of the local contributions of the inhabitants and adjoining proprietors to a considerably greater extent; the result of which was a great increase in the shipping accommodation of the coast towns, to the benefit of the local population, and of shipowners and navigators generally.

Mr. Telford's principal harbour works in Scotland, however, were those of Aberdeen and Dundee, which, next to Leith (the port of Edinburgh), formed the principal havens along the east coast. The neighbourhood of Aberdeen was originally so wild and barren that Telford expressed his surprise that any class of men should ever have settled there. An immense shoulder of the Grampian mountains extends down to the sea-coast, where it terminates in a bold, rude promontory. The country on either side of the Dee, which flows past the town, was originally covered with innumerable granite blocks; one, called Craig Metellan, lying right in the river's mouth, and forming, with the sand, an almost effectual bar to its navigation. Although, in ancient times, a little cultivable land lay immediately outside the town, the region beyond was as sterile as it is possible for land to be in such a latitude. "Any wher," says an ancient writer, "after yow pass a myll without the toune, the countrey is barren lyke, the hills craigy, the plaines full of marishes and mosses, the feilds are covered with heather or peeble stons, the corne feilds mixt with thes bot few. The air is temperat and healthful about it, and it may be that the citizens owe the acuteness of their wits thereunto and their civill inclinations; the lyke not easie to be found under northerlie climats, damped for the most pairt with air of a grosse consistence." *

But the old inhabitants of Aberdeen and its neighbourhood were really as rough as their soil. Judged by their records, they must have been dreadfully haunted by witches and sorcerers down to a comparatively recent period; witch-burning having been common in the town until the end of the sixteenth century. We find that, in one year, no fewer than twenty-three women and one man were burnt; the Dean of Guild Records containing the detailed accounts of the "loads of peattis, tar barrellis," and other combustibles used in burning them. The lairds of the Garioch, a district

* 'A Description of Bothe Touns of Aberdeene.' By James Gordon, Parson of Rothiemay. Reprinted in Gavin Turreff's 'Antiquarian Gleanings from Aberdeenshire Records.' Aberdeen, 1859.

in the immediate neighbourhood, seem to have been still more terrible than the witches, being accustomed to enter the place and make an onslaught upon the citizens, according as local rage and thirst for spoil might incline them. On one of such occasions, eighty of the inhabitants were killed and wounded.* Down even to the middle of last century the Aberdonian notions of personal liberty seem to have been very restricted; for between 1740 and 1746 we find that persons of both sexes were kidnapped, put on board ships, and despatched to the American plantations, where they were sold for slaves. Strangest of all, the men who carried on this slave trade were local dignitaries, one of them being a town's baillie, another the town-clerk depute. Those kidnapped were openly "driven in flocks through the town, like herds of sheep, under the care of a keeper armed with a whip."† So open was the traffic that the public workhouse was used for their reception until the ships sailed, and when that was filled, the tolbooth or common prison was made use of. The vessels which sailed from the harbour for America in 1743 contained no fewer than sixty-nine persons; and it is supposed that, in the six years during which the Aberdeen slave trade was at its height, about six hundred were transported for sale, very few of whom ever returned.‡

This slave traffic was doubtless stimulated by the foreign

---

* Robertson's 'Book of Bon-Accord.'

† Ibid., quoted in Turreff's 'Antiquarian Gleanings,' p. 222.

‡ One of them, however, did return—Peter Williamson, a native of the town, sold for a slave in Pennsylvania, "a rough, ragged, humle-headed, long, stowie, clever boy," who, reaching York, published an account of the infamous traffic, in a pamphlet which excited extraordinary interest at the time, and met with a rapid and extensive circulation. But his exposure of kidnapping gave very great offence to the magistrates, who dragged him before their tribunal as having "published a scurrilous and infamous libel on the corporation," and he was sentenced to be imprisoned until he should sign a denial of the truth of his statements. He brought an action against the corporation for their proceedings, and obtained a verdict and damages; and he further proceeded against Baillie Fordyce (one of his kidnappers), and others, from whom he obtained 200*l.* damages, with costs. The system was thus effectually put a stop to.

ships beginning to frequent the port; for the inhabitants were industrious, and their plaiding, linen, and worsted stockings were in much request as articles of merchandise. Cured salmon were also exported in large quantities. As early as 1659, a quay was formed along the Dee towards the village of Foot Dee. "Beyond Futty," says an old writer, "lyes the fisher-boat heavne; and after that, towards the promontorie called Sandenesse, ther is to be seen a grosse bulk of a building, vaulted and flatted above (the Blockhous they call it), begun to be builded anno 1513, for guarding the entree of the harboree from pirats and algarads; and cannon wer planted ther for that purpose, or, at least, that from thence the motions of pirats might be tymouslie foreseen. This rough piece of work was finished anno 1542, in which yer lykewayes the mouth of the river Dee was locked with cheans of iron and masts of ships crossing the river, not to be opened bot at the citizens' pleasure." *

After the Union, but more especially after the rebellion of 1745, the trade of Aberdeen made considerable progress. Although Burns, in 1787, briefly described the place as a "lazy toun," the inhabitants were displaying much energy in carrying out improvements in their port.† In 1775 the foundation-stone of the new pier designed by Mr. Smeaton was laid with great ceremony, and, the works proceeding to completion, a new pier, twelve hundred feet long, terminating in a round head, was finished in less than six years. The trade of the place was, however, as yet too small to justify anything beyond a tidal harbour, and the

* 'A Description of Bothe Touns of Aberdeene.' By James Gordon, Parson of Rothiemay. Quoted by Turreff, p. 109.

† Communication with London was as yet by no means frequent, and far from expeditious, as the following advertisement of 1778 will show:—"For London: To sail positively on Saturday next, the 7th November, wind and weather permitting, the *Aberdeen* smack. Will lie a short time at London, and, if no convoy is appointed, will sail under care of a fleet of colliers —the best convoy of any. For particulars apply," &c., &c.

engineer's views were limited to that object. He found the river meandering over an irregular space about five hundred yards in breadth; and he applied the only practicable remedy, by confining the channel as much as the limited means placed at his disposal enabled him to do, and directing the land floods so as to act upon and diminish the bar. Opposite the north pier, on the south side of the river, Smeaton constructed a breast-wall about half the length of the pier. Owing, however, to a departure from that engineer's plans, by which the pier was placed too far to the north, it was found that a heavy swell entered the harbour, and, to obviate this formidable inconvenience, a bulwark was projected from it, so as to occupy about one third of the channel entrance.

The trade of the place continuing to increase, Mr. Rennie was called upon, in 1797, to examine and report upon the best means of improving the harbour, when he recommended the construction of floating docks upon the sandy flats called Foot Dee. Nothing was done at the time, as the scheme was very costly and considered beyond the available means of the locality. But the magistrates kept the subject in mind; and when Mr. Telford made his report on the best means of improving the harbour in 1801, he intimated that the inhabitants were ready to cooperate with the Government in rendering it capable of accommodating ships of war, as far as their circumstances would permit.

In 1807, the south pier-head, built by Smeaton, was destroyed by a storm, and the time had arrived when something must be done, not only to improve but even to preserve the port. The magistrates accordingly proceeded, in 1809, to rebuild the pier-head of cut granite, and at the same time they applied to Parliament for authority to carry out further improvements after the plan recommended by Mr. Telford; and the necessary powers were conferred in the following year. The new works comprehended a large extension of the wharfage accommodation, the construction of floating and graving docks, increased means of scouring the harbour and ensuring greater depth of water on the bar across the river's mouth, and the provision of a navigable communication between the Aberdeenshire Canal and the new harbour.

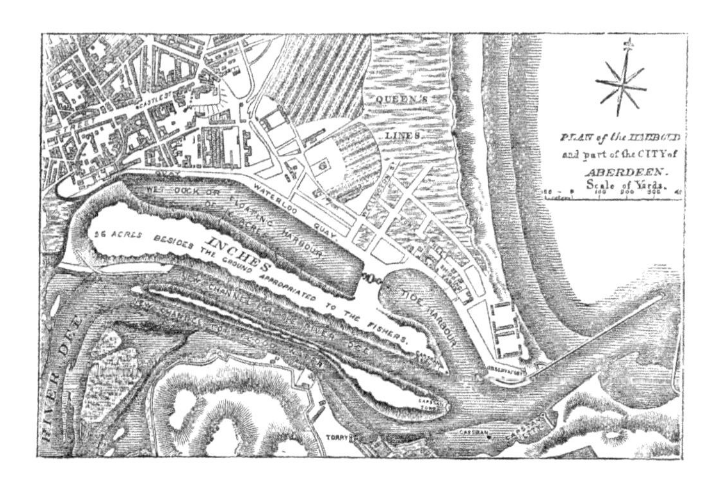

The extension of the north pier was first proceeded with, under the superintendence of John Gibb, the resident engineer; and by the year 1811 the whole length of 300 additional feet had been completed. The beneficial effects of this extension were so apparent, that a general wish was expressed that it should be carried further; and it was eventually determined to extend the pier 780 feet beyond Smeaton's head, by which not only was much deeper water secured, but vessels were better enabled to clear the Girdleness Point. This extension was successfully carried out by the end of the year 1812. A strong breakwater, about 800 feet long, was also run out from the south shore, leaving a space of about 250 feet as an entrance, thereby giving greater protection to the shipping in the harbour, while the contraction of the channel, by increasing the

ABERDEEN HARBOUR.

By R. P LEITCH.

Page 222.

"scour," tended to give a much greater depth of water on the bar.

The outer head of the pier was seriously injured by the heavy storms of the two succeeding winters, which rendered it necessary to alter its formation to a very flat slope of about five to one all round the head.* New wharves were at the same time constructed inside the harbour; a new channel for the river was excavated, which further enlarged the floating space and wharf accommodation; wet and dry docks were added; until at length the quay berthage amounted to not less than 6290 feet, or nearly a mile and a quarter in length. By these combined improvements an additional extent of quay room was obtained of

* "The bottom under the foundations," says Mr. Gibb, in his description of the work, "is nothing better than loose sand and gravel, constantly thrown up by the sea on that stormy coast, so that it was necessary to consolidate the work under low water by dropping large stones from lighters, and filling the interstices with smaller ones, until it was brought within about a foot of the level of low water, when the ashlar work was commenced; but in place of laying the stones horizontally in their beds, each course was laid at an angle of 45 degrees, to within about 18 inches of the top, when a level coping was added. This mode of building enabled the work to be carried on expeditiously, and rendered it while in progress less liable to temporary damage, likewise affording three points of bearing; for while the ashlar walling was carrying up on both sides, the middle or body of the pier was carried up at the same time by a careful backing throughout of large rubble-stone, to within 18 inches of the top, when the whole was covered with granite coping and paving 18 inches deep, with a cut granite parapet wall on the north side of the whole length of the pier, thus protected for the convenience of those who might have occasion to frequent it." — Mr. Gibb's 'Narrative of Aberdeen Harbour Works.'

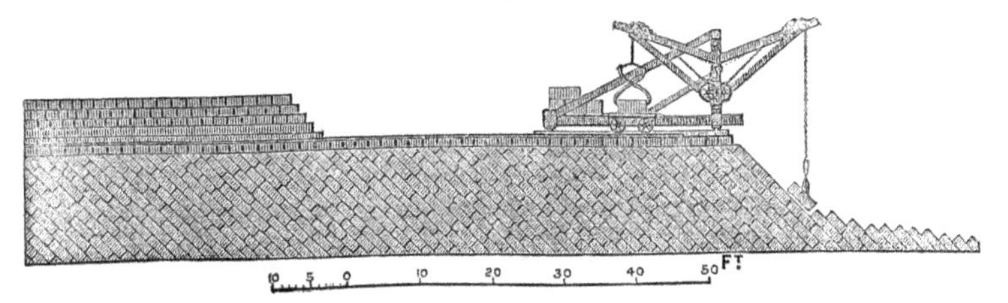

SECTION OF PIER-HEAD WORK.

about 4000 feet; an excellent tidal harbour was formed, in which, at spring tides, the depth of water is about 15 feet; while on the bar it was increased to about 19 feet. The prosperity of Aberdeen had meanwhile been advancing apace. The city had been greatly beautified and enlarged: shipbuilding had made rapid progress; Aberdeen clippers became famous, and Aberdeen merchants carried on a trade with all parts of the world; manufactures of wool, cotton, flax, and iron were carried on with great success; its population rapidly increased; and, as a maritime city, Aberdeen took rank as the third in Scotland, the tonnage entering the port having increased from 50,000 tons in 1800 to about 300,000 in 1860.

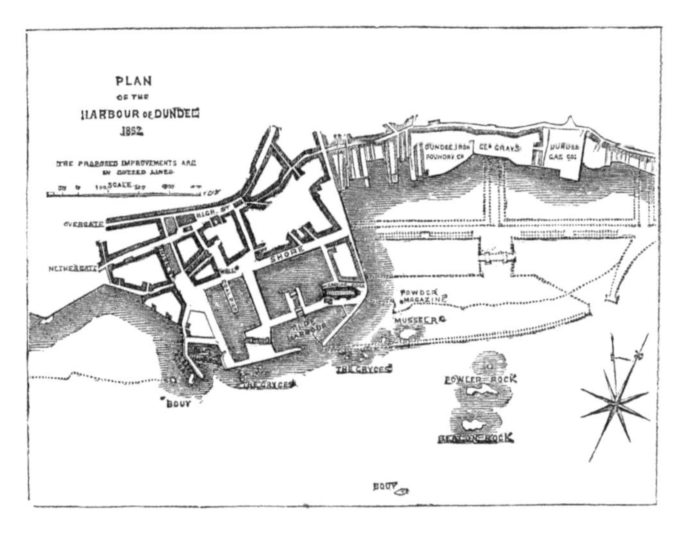

Improvements of an equally important character were carried out by Mr. Telford in the port of Dundee, also situated on the east coast of Scotland, at the entrance to the Frith of Tay. There are those still living at the place who remember its former haven, consisting of a crooked wall, affording shelter to only a few fishing-boats or smuggling vessels — its trade being then altogether paltry, scarcely deserving the name, and its population not one-fifth of what it now is. Helped by its commodious and capacious harbour, it has become one of the most populous and thriving towns on the east coast.

DUNDEE HARBOUR. [By R. P. Leitch.]

The trade of the place took a great start forward at the close of the war, and Mr. Telford was called upon to supply the plans of a new harbour. His first design, which he submitted in 1814, was of a comparatively limited character; but it was greatly enlarged during the progress of the works. Floating docks were added, as well as graving docks for large vessels. The necessary powers were obtained in 1815; the works proceeded vigorously under the Harbour Commissioners, who superseded the old obstructive corporation; and in 1825 the splendid new floating dock—750 feet long by 450 broad, having an entrance-lock 170 feet long and 40 feet wide—was opened to the shipping of all countries.

CHAPTER X.

CALEDONIAN AND OTHER CANALS.

THE formation of a navigable highway through the chain of locks lying in the Great Glen of the Highlands, and extending diagonally across Scotland from the Atlantic to the North Sea, had long been regarded as a work of national importance. As early as 1773, James Watt, then following the business of a land-surveyor at Glasgow, made a survey of the country at the instance of the Commissioners of Forfeited Estates. He pronounced the canal practicable, and pointed out how it could best be constructed. There was certainly no want of water, for Watt was repeatedly drenched with rain while he was making his survey, and he had difficulty in preserving even his journal book. "On my way home," he says, "I passed through the wildest country I ever saw, and over the worst conducted roads."

Twenty years later, in 1793, Mr. Rennie was consulted as to the canal, and he also prepared a scheme: but nothing was done. The project was, however, revived in 1801 during the war with Napoleon, when various inland ship canals—such as those from London to Portsmouth, and from Bristol to the English Channel—were under consideration with the view of enabling British shipping to pass from one part of the kingdom to another without being exposed to the attacks of French privateers. But there was another reason for urging the formation of the canal through the Great Glen of Scotland, which was regarded as of considerable importance before the introduction of steam enabled vessels to set the winds and tides at comparative defiance. It was this: vessels sailing from the eastern ports to America had to beat up the Pentland Frith, often against adverse winds and stormy seas, which rendered the navigation both tedious and dangerous. Thus it was cited by Sir Edward Parry, in his evidence before Parliament in favour of completing the Caledonian Canal, that of two vessels despatched from Newcastle on the same day—one bound for Liverpool by the north of Scotland, and the other for Bombay by the English Channel and the Cape of Good Hope—the latter reached its destination first! Another case may be mentioned, that of an Inverness vessel, which sailed for Liverpool on a Christmas Day, reached Stromness Harbour, in Orkney, on the 1st of January, and lay there windbound, with a fleet of other traders, until the middle of April following! In fact, the Pentland Frith, which is the throat connecting the Atlantic and German Oceans, through which the former rolls its long majestic waves with tremendous force, was long the dread of mariners, and it was considered an object of national importance to mitigate the dangers of the passage towards the western seas.

As the lochs occupying the chief part of the bottom of the Great Glen were of sufficient depth to be navigable by large vessels, it was thought that if they could be connected by a ship canal, so as to render the line of navigation continuous, it would be used by shipping to a large extent, and prove of great public service. Five hundred miles of dangerous navigation by the Orkneys and Cape Wrath would thereby be saved, while ships of war, were this track open to them, might reach the north of Ireland in two days from Fort George near Inverness.

When the scheme of the proposed canal was revived in 1801, Mr. Telford was requested to make a survey and send in his report on the subject. He immediately wrote to his friend James Watt, saying, "I have so long accustomed myself to look with a degree of reverence at your work, that I am particularly anxious to learn what occurred to you in this business while the whole was fresh in your mind. The object appears to me so great and so desirable,

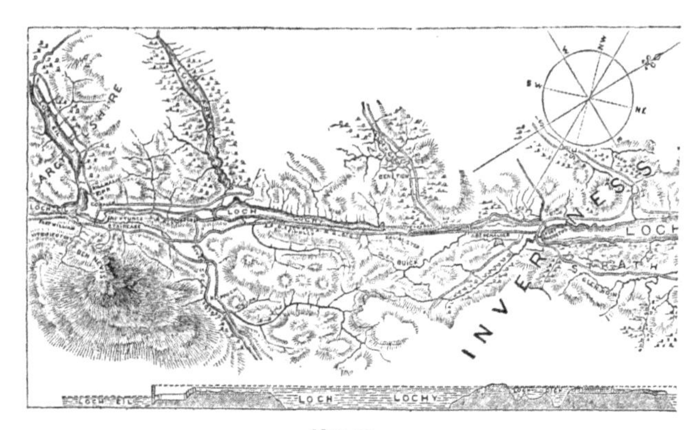

MAP OF

that I am convinced you will feel a pleasure in bringing it again under investigation, and I am very desirous that the thing should be fully and fairly explained, so that the public may be made aware of its extensive utility. If I can accomplish this, I shall have done my duty; and if the project is not executed now, some future period will see it done, and I shall have the satisfaction of having followed you and promoted its success." We may here state that Telford's survey agreed with Watt's in the most important particulars, and that he largely cited Watt's descriptions of the proposed scheme in his own report.

Mr. Telford's first inspection of the district was made in 1801, and his report was sent in to the Treasury in the course of the following year. Lord Bexley, then Secretary to the Treasury, took a warm personal interest in the project, and lost no opportunity of actively promoting it. A board of commissioners was eventually appointed to carry out the formation of the canal. Mr. Telford, on being appointed principal engineer of the undertaking, was requested at once to proceed to Scotland and prepare the necessary working survey. He was accompanied on the occasion by Mr. Jessop as consulting engineer. Twenty thousand pounds were granted under the provisions of the

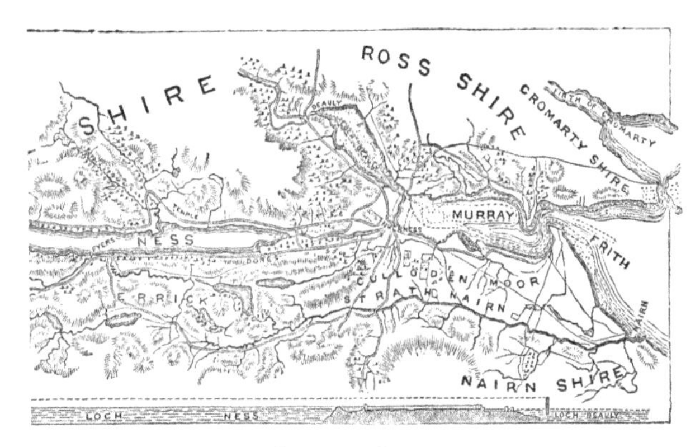

CALEDONIAN CANAL.

43 Geo. III. (chap. cii.), and the works were commenced, in the beginning of 1804, by the formation of a dock or basin adjoining the intended tide-lock at Corpach, near Bannavie.

The basin at Corpach formed the southernmost point of the intended canal. It is situated at the head of Loch Eil, amidst some of the grandest scenery of the Highlands. Across the Loch is the little town of Fort William, one of the forts established at the end of the seventeenth century to keep the wild Highlanders in subjection. Above it rise hills over hills, of all forms and sizes, and of all hues, from grass-green below to heather-brown and purple above, capped with heights of weather-beaten grey; while towering over all stands the rugged mass of Ben Nevis—a mountain almost unsurpassed for picturesque grandeur. Along the western foot of the range, which extends for some six or eight miles, lies a long extent of brown bog, on the verge of which, by the river Lochy, stand the ruins of Inverlochy Castle.

The works at Corpach involved great labour, and extended over a long series of years. The difference between the level of Loch Eil and Loch Lochy is ninety feet, while the distance between them was less than eight miles.

It was therefore necessary to climb up the side of the hill by a flight of eight gigantic locks, clustered together, and which Telford named Neptune's Staircase. The ground passed over was in some places very difficult, requiring large masses of embankment, the slips of which in the course of the work frequently occasioned serious embarrassment. The basin on Loch Eil, on the other hand, was constructed amidst rock, and considerable difficulty was experienced in getting in the necessary coffer-dam for the construction of the opening into the sea-lock, the entrance-sill of which was laid upon the rock itself, so that there was a depth of 21 feet of water upon it at high water of neap tides.

At the same time that the works at Corpach were begun, the dock or basin at the north-eastern extremity of the canal, situated at Clachnaharry, on the shore of Loch Beauly, was also laid out, and the excavations and embankments were carried on with considerable activity. This dock was constructed about 967 yards long, and upwards of 162 yards in breadth, giving an area of about 32 acres,—forming, in fact, a harbour for the vessels using the canal. The dimensions of the artificial waterway were of unusual size, as the intention was to adapt it throughout for the passage of a 32-gun frigate of that day, fully equipped and laden with stores. The canal, as originally resolved upon, was designed to be 110 feet wide at the surface, and 50 feet at the bottom, with a depth in the middle of 20 feet; though these dimensions were somewhat modified in the execution of the work. The locks were of corresponding large dimensions, each being from 170 to 180 feet long, 40 broad, and 20 deep.

Between these two extremities of the canal—Corpach on the south-west and Clachnaharry on the north-east—extends the chain of fresh-water lochs: Loch Lochy on the south; next Loch Oich; then Loch Ness; and lastly, furthest north, the small Loch of Dochfour. The whole length of the navigation is 60 miles 40 chains, of which the navigable lochs constitute about 40 miles, leaving only about 20 miles of canal to be constructed, but of unusually large dimensions and through a very difficult country.

The summit loch of the whole is Loch Oich, the surface of which is exactly a hundred feet above high water-mark, both at Inverness and Fort William; and to this sheet of water the navigation climbs up by a series of locks from both the eastern and western seas. The whole number of these is twenty-eight: the entrance-lock at Clachnaharry, constructed on piles, at the end of huge embankments, forced out into deep water, at Loch Beauly; another at the en-

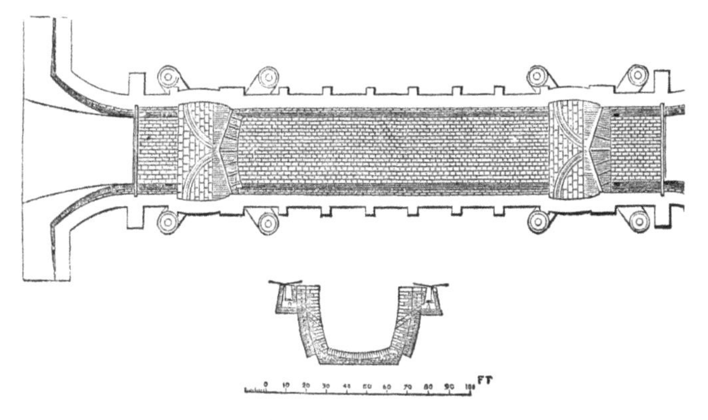

LOCK, CALEDONIAN CANAL.

trance to the capacious artificial harbour above mentioned, at Muirtown; four connected locks at the southern end of this basin; a regulating lock a little to the north of Loch Dochfour; five contiguous locks at Fort Augustus, at the south end of Loch Ness; another, called the Kytra Lock, about midway between Fort Augustus and Loch Oich; a regulating lock at the north-east end of Loch Oich; two contiguous locks between Lochs Oich and Lochy; a regulating lock at the south-west end of Loch Lochy; next, the grand series of locks, eight in number, called "Neptune's

Staircase," at Bannavie, within a mile and a quarter of the sea; two locks, descending to Corpach basin; and lastly, the great entrance or sea-lock at Corpach.

The northern entrance-lock from the sea at Loch Beauly is at Clachnaharry, near Inverness. The works here were not accomplished without much difficulty as well as labour, partly from the very gradual declivity of the shore, and partly from the necessity of placing the sea-lock on absolute mud, which afforded no foundation other than what was created by compression and pile-driving. The mud was forced down by throwing upon it an immense load of earth and stones, which was left during twelve months to settle; after which a shaft was sunk to a solid foundation, and the masonry of the sea-lock was then founded and built therein.

In the 'Sixteenth Report of the Commissioners of the Caledonian Canal,' the following reference is made to this important work, which was finished in 1812:—"The depth of the mud on which it may be said to be artificially seated is not less than 60 feet; so that it cannot be deemed superfluous, at the end of seven years, to state that no subsidence is discoverable; and we presume that the entire lock, as well as every part of it, may now be deemed as immovable, and as little liable to destruction, as any other large mass of masonry. This was the most remarkable work performed under the immediate care of Mr. Matthew Davidson, our superintendent at Clachnaharry, from 1804 till the time of his decease. He was a man perfectly qualified for the employment by inflexible integrity, unwearied industry, and zeal to a degree of anxiety, in all the operations committed to his care."*

* Mr. Matthew Davidson, above referred to, was an excellent officer, but a strange cynical humourist in his way. He was a Lowlander, and had lived for some time in England, at the Pont Cysylltau works, where he had acquired a taste for English comforts, and returned to the North with a considerable contempt for the Highland people amongst whom he was stationed. He is said to have very much resembled Dr. Johnson in person, and was so fond of books,

As may naturally be supposed, the execution of these great works involved vast labour and anxiety. They were designed with much skill, and executed with equal ability. There were lock-gates to be constructed, principally of cast iron, sheathed with pine planking. Eight public road bridges crossed the line of the canal, which were made of cast iron, and swung horizontally. There were many mountain streams, swollen to torrents in winter, crossing under the canal, for which abundant water-way had to be provided, involving the construction of numerous culverts, tunnels, and under-bridges of large dimensions. There were also powerful sluices to let off the excess of water sent down from the adjacent mountains into the canal during winter. Three of these, of great size, high above the river Lochy, are constructed at a point where the canal is cut through the solid rock; and the sight of the mass of waters rushing down into the valley beneath, gives an impression of power which, once seen, is never forgotten.

These great works were only brought to a completion after the labours of many years, during which the difficulties encountered in their construction had swelled the cost of the canal far beyond the original estimate. The rapid advances which had taken place in the interval in the prices of labour and materials also tended greatly to increase the expenses, and, after all, the canal, when completed and opened, was comparatively little used. This was doubtless owing, in a great measure, to the rapid changes which occurred in the system of navigation shortly after the projection of the undertaking. For these Telford

and so well read in them, that he was called 'the Walking Library.' He used to say that if justice were done to the inhabitants of Inverness, there would be nobody left there in twenty years but the Provost and the hangman. Seeing an artist one day making a sketch in the mountains, he said it was the first time he had known what the hills were good for. And when some one was complaining of the weather in the Highlands, he looked sarcastically round, and observed that the rain certainly would not hurt the heather crop.

was not responsible. He was called upon to make the canal, and he did so in the best manner Engineers are not required to speculate as to the commercial value of the works they are required to construct; and there were circumstances connected with the scheme of the Caledonian Canal which removed it from the category of mere commercial adventures. It was a Government project, and it proved a failure as a paying concern. Hence it formed a prominent topic for discussion in the journals of the day; but the attacks made upon the Government because of their expenditure on the hapless undertaking were perhaps more felt by Telford, who was its engineer, than by all the ministers of state conjoined.

"The unfortunate issue of this great work," writes the present engineer of the canal, to whom we are indebted for many of the preceding facts, "was a grievous disappointment to Mr. Telford, and was in fact the one great bitter in his otherwise unalloyed cup of happiness and prosperity. The undertaking was maligned by thousands who knew nothing of its character. It became 'a dog with a bad name,' and all the proverbial consequences followed. The most absurd errors and misconceptions were propagated respecting it from year to year, and it was impossible during Telford's lifetime to stem the torrent of popular prejudice and objurgation. It must, however, be admitted, after a long experience, that Telford was greatly over-sanguine in his expectations as to the national uses of the canal, and he was doomed to suffer acutely in his personal feelings, little though he may have been personally to blame, the consequences of what in this commercial country is regarded as so much worse than a crime, namely, a financial mistake."*

* The misfortunes of the Caledonian Canal did not end with the life of Telford. The first vessel passed through it from sea to sea in October, 1822, by which time it had cost about a million sterling, or double the original estimate. Notwithstanding this large outlay, it appears that the canal was opened before the works had been

Mr. Telford's great sensitiveness made him feel the ill success of this enterprise far more than most other men would have done. He was accustomed to throw himself into the projects on which he was employed with an enthusiasm almost poetic. He regarded them not merely as so much engineering, but as works which were to be instrumental in opening up the communications of the country and extending its civilization. Viewed in this light, his canals, roads, bridges, and harbours were unquestionably of great national importance, though their commercial results might not in all cases justify the estimates of their projectors. To refer to like instances—no one can doubt the immense value and public uses of Mr. Rennie's Waterloo Bridge or Mr. Robert Stephenson's Britannia and Victoria Bridges, though every one knows that, commercially, they have been failures. But it is probable that neither of these eminent engineers gave himself anything like the anxious concern that Telford did about the financial issue of his undertaking. Were railway engineers to fret and vex themselves about the commercial value of the schemes in which they have been engaged, there are few of them but would be so haunted by the ghosts of wrecked speculations that they could scarcely lay their heads upon their pillows for a single night in peace.

While the Caledonian Canal was in progress, Mr. Telford was occupied in various works of a similar kind in England and Scotland, and also upon one in Sweden. In 1804, while

properly completed; and the consequence was that they very shortly fell into decay. It even began to be considered whether the canal ought not to be abandoned. In 1838 Mr. James Walker, C.E., an engineer of the highest eminence, examined it, and reported fully on its then state, strongly recommending its completion as well as its improvement. His advice was eventually adopted, and the canal was finished accordingly, at an additional cost of about 200,000*l.*, and the whole line was re-opened in 1847, since which time it has continued in useful operation. The passage from sea to sea at all times can now be depended on, and it can usually be made in forty-eight hours. As the trade of the North increases, the uses of the canal will probably become much more decided than they have heretofore proved.

on one of his journeys to the north, he was requested by the Earl of Eglinton and others to examine a project for making a canal from Glasgow to Saltcoats and Ardrossan, on the north-western coast of the county of Ayr, passing near the important manufacturing town of Paisley. A new survey of the line was made, and the works were carried on during several successive years until a very fine capacious canal was completed, on the same level, as far as Paisley and Johnstown. But the funds of the company falling short, the works were stopped, and the canal was carried no further. Besides, the measures adopted by the Clyde Trustees to deepen the bed of that river and enable ships of large burden to pass up as high as Glasgow, had proved so successful that the ultimate extension of the canal to Ardrossan was no longer deemed necessary, and the prosecution of the work was accordingly abandoned. But as Mr. Telford has observed, no person suspected, when the canal was laid out in 1805, "that steamboats would not only monopolise the trade of the Clyde, but penetrate into every creek where there is water to float them, in the British Isles and the continent of Europe, and be seen in every quarter of the world."

Another of the navigations on which Mr. Telford was long employed was that of the river Weaver in Cheshire. It was only twenty-four miles in extent, but of considerable importance to the country through which it passed, accommodating the salt-manufacturing districts, of which the towns of Nantwich, Northwich, and Frodsham are the centres. The channel of the river was extremely crooked and much obstructed by shoals, when Telford took the navigation in hand in the year 1807, and a number of essential improvements were made in it, by means of new locks, weirs, and side cuts, which had the effect of greatly improving the communications of these important districts.

In the following year we find our engineer consulted, at the instance of the King of Sweden, on the best mode of constructing the Gotha Canal, between Lake Wenern and the Baltic, to complete the communication with the North Sea. In 1808, at the invitation of Count Platen, Mr. Telford visited Sweden and made a careful survey of the district. The service occupied him and his assistants two months, after which he prepared and sent in a series of detailed plans and sections, together with an elaborate report on the subject. His plans having been adopted, he again visited Sweden in 1810, to inspect the excavations which had already been begun, when he supplied the drawings for the locks and bridges. With the sanction of the British Government, he at the same time furnished the Swedish contractors with patterns of the most improved tools used in canal making, and took with him a number of experienced lock-makers and navvies for the purpose of instructing the native workmen.

The construction of the Gotha Canal was an undertaking of great magnitude and difficulty, similar in many respects to the Caledonian Canal, though much more extensive. The length of artificial canal was 55 miles, and of the whole navigation, including the lakes, 120 miles. The locks are 120 feet long and 24 feet broad; the width of the canal at bottom being 42 feet, and the depth of water 10 feet. The results, so far as the engineer was concerned, were much more satisfactory than in the case of the Caledonian Canal. While in the one case he had much obloquy to suffer for the services he had given, in the other he was honoured and fêted as a public benefactor, the King conferring upon him the Swedish order of knighthood, and presenting him with his portrait set in diamonds.

Among the various canals throughout England which Mr. Telford was employed to construct or improve, down to the commencement of the railway era, were the Gloucester and Berkeley Canal, in 1818; the Grand Trunk Canal, in 1822; the Harecastle Tunnel, which he constructed anew, in 1824-7; the Birmingham Canal, in 1824; and the Macclesfield, and Birmingham and Liverpool Junction Canals, in 1825. The Gloucester and Berkeley Canal Company had been unable to finish their works, begun some thirty years before; but with the assistance of a loan of 160,000*l.* from the Exchequer Bill Loan Commissioners, they were enabled to proceed with the completion of their undertaking. A capacious canal was cut from Gloucester to Sharpness Point, about eight miles down the Severn, which had the effect of greatly improving the convenience of the port of Gloucester; and by means of this navigation, ships of large burden can now avoid the circuitous and difficult passage of the higher part of the river, very much to the advantage of the trade of the place.

The formation of a new tunnel through Harecastle Hill, for the better accommodation of the boats passing along the Grand Trunk Canal, was a formidable work. The original tunnel, it will be remembered,* was laid out by Brindley, about fifty years before, and occupied eleven years in construction. But the engineering appliances of those early days were very limited; the pumping powers of the steam-engine had not been fairly developed, and workmen were as yet only half-educated in the expert use of tools. The tunnel, no doubt, answered the purpose for which it was originally intended, but it was very soon found too limited for the traffic passing along the navigation. It was little larger than a sewer, and admitted the passage of only one narrow boat, seven feet wide, at a time, involving very heavy labour on the part of the men who worked it through. This was performed by what was called *legging.* The Leggers lay upon the deck of the vessel, or upon a board slightly projecting from either side of it, and, by thrusting their feet against the slimy roof or sides of the tunnel—walking horizontally as it were—they contrived to push it through. But it was no better than horse-work; and after "legging" Harecastle Tunnel, which is more than a mile and a half long, the men were usually

* 'Brindley and the Early Engineers,' p. 267.

completely exhausted, and as wet from perspiration as if they had been dragged through the canal itself. The process occupied about two hours, and by the time the passage of the tunnel was made, there was usually a collection of boats at the other end waiting their turn to pass. Thus much contention and confusion took place amongst the boatmen—a very rough class of labourers—and many furious battles were fought by the claimants for the first turn "through." Regulations were found of no avail to settle these disputes, still less to accommodate the large traffic which continued to keep flowing along the line of the Grand Trunk, and steadily increased with the advancing trade and manufactures of the country. Loud complaints were made by the public, but they were disregarded for many years; and it was not until the proprietors were threatened with rival canals and railroads that they determined on—what they could no longer avoid if they desired to retain the carrying trade of the district—the enlargement of the Harecastle Tunnel.

Mr. Telford was requested to advise the Company what course was most proper to be adopted in the matter, and after examining the place, he recommended that an entirely new tunnel should be constructed, nearly parallel with the old one, but of much larger dimensions. The work was begun in 1824, and completed in 1827, in less than three years. There were at that time throughout the country plenty of skilled labourers and contractors, many of them trained by their experience upon Telford's own works, whereas Brindley had in a great measure to make his workmen out of the rawest material. Telford also had the advantage of greatly improved machinery and an abundant supply of money—the Grand Trunk Canal Company having become prosperous and rich, paying large dividends. It is therefore meet, while eulogising the despatch with which he was enabled to carry out the work, to point out that the much greater period occupied in the earlier undertaking is not to be set down to the disparagement of

Brindley, who had difficulties to encounter which the later engineer knew nothing of.

The length of the new tunnel is 2926 yards; it is 16 feet high and 14 feet broad, 4 feet 9 inches of the breadth being occupied by the towing-path—for "legging" was now dispensed with, and horses hauled along the boats instead of their being thrust through by men. The tunnel is in so perfectly straight a line that its whole length can be seen through at one view; and though it was constructed by means of fifteen different pitshafts sunk to the same line along the length of the tunnel, the workmanship is so perfect that the joinings of the various lengths of brickwork are scarcely discernible. The convenience afforded by the new tunnel was very great, and Telford mentions that, on surveying it in 1829, he asked a boatman coming out of it how he liked it? "I only wish," he replied, "that it reached all the way to Manchester!"

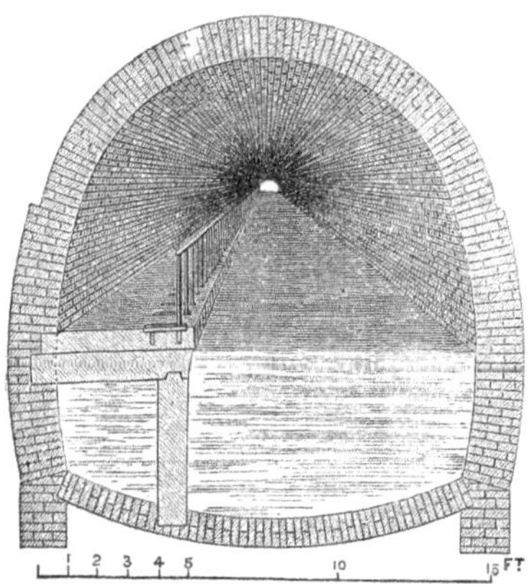

CROSS SECTION OF HARECASTLE TUNNEL.

At the time that Mr. Telford was engaged upon the tunnel at Harecastle, he was employed to improve and widen the Birmingham Canal, another of Brindley's works. Though the accommodation provided by it had been sufficient for the traffic when originally constructed, the expansion of the trade of Birmingham and the neighbourhood, accelerated by the formation of the canal itself, had been such as completely to outgrow its limited convenience and capacity, and its enlargement and improvement now became absolutely necessary. Brindley's Canal, for the sake of cheapness of construction—money being much scarcer and more difficult to be raised in the early days of canals—was also winding and crooked; and it was considered desirable to shorten and straighten it by cutting off the bends at different places. At the point at which the canal entered Birmingham, it had become "little better than a crooked ditch, with scarcely the appearance of a towing-path, the horses frequently sliding and staggering in the water, the hauling-lines sweeping the gravel into the canal, and the entanglement at the meeting of boats being incessant; whilst at the locks at each end of the short summit at Smethwick crowds of boatmen were always quarrelling, or offering premiums for a preference of passage; and the mine-owners, injured by the delay, were loud in their just complaints."*

Mr. Telford proposed an effective measure of improvement, which was taken in hand without loss of time, and carried out, greatly to the advantage of the trade of the district. The numerous bends in the canal were cut off, the water-way was greatly widened, the summit at Smethwick was cut down to the level on either side, and a straight canal, forty feet wide, without a lock, was thus formed as far as Bilston and Wolverhampton; while the length of the main line between Birmingham and Autherley, along the whole extent of the "Black country," was reduced from twenty-two to fourteen miles. At the same time the obsolete curvatures in Brindley's old canal were converted into separate branches or basins, for the accommodation of the numerous mines and manufactories on either side of the main line. In consequence of the alterations which had been made in the canal, it was found necessary to construct numerous large bridges. One of these—a cast iron bridge, at Galton, of 150 feet span—has been much admired for its elegance, lightness, and economy of material. Several others of cast iron were constructed at different points, and at one place the canal itself is carried along on an aqueduct

* 'Life of Telford,' p. 82, 83.

of the same material as at Pont-Cysylltau. The whole of these extensive improvements were carried out in the short space of two years; and the result was highly satisfactory, "proving," as Mr. Telford himself observes, "that where business is extensive, liberal expenditure of this kind is true economy."

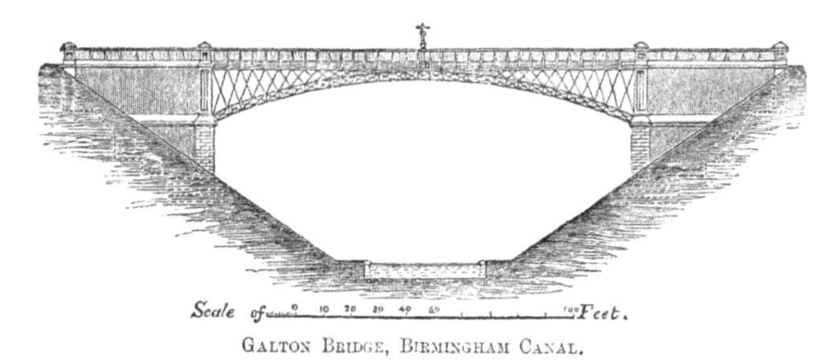

GALTON BRIDGE, BIRMINGHAM CANAL.

In 1825 Mr. Telford was called upon to lay out a canal to connect the Grand Trunk, at the north end of Harecastle Tunnel, with the rapidly improving towns of Congleton and Macclesfield. The line was twenty-nine miles in length, ten miles on one level from Harecastle to beyond Congleton; then, ascending 114 feet by eleven locks, it proceeded for five miles on a level past Macclesfield, and onward to join the Peak Forest Canal at Marple. The navigation was thus conducted upon two levels, each of considerable length; and it so happened that the trade of each was in a measure distinct, and required separate accommodation. The traffic of the whole of the Congleton district had ready access to the Grand Trunk system, without the labour, expense, and delay involved by passing the boats through locks; while the coals brought to Macclesfield to supply the mills there were carried throughout upon the upper level, also without lockage. The engineer's arrangement proved highly judicious, and furnishes an illustration of the tact and judgment which he usually displayed in laying out his works for practical uses. Mr.

Telford largely employed cast iron in the construction of this canal, using it in the locks and gates, as well as in an extensive aqueduct which it was necessary to construct over a deep ravine, after the plan pursued by him at Pont-Cysylltau and other places.

The last canal constructed by Mr. Telford was the Birmingham and Liverpool Junction, extending from the Birmingham Canal, near Wolverhampton, in nearly a direct line, by Market Drayton, Nantwich, and through the city of Chester, by the Ellesmere Canal, to Ellesmere Port on the Mersey. The proprietors of canals were becoming alarmed at the numerous railways projected through the districts heretofore served by their water-ways; and among other projects one was set on foot, as early as 1825, for constructing a line of railway from London to Liverpool. Mr. Telford was consulted as to the best means of protecting existing investments, and his advice was to render the canal system as complete as it could be made; for he entertained the conviction, which has been justified by experience, that such navigations possessed peculiar advantages for the conveyance of heavy goods, and that, if the interruptions presented by locks could be done away with, or materially reduced, a large portion of the trade of the country must continue to be carried by the water roads. The new line recommended by him was approved and adopted, and the works were commenced in 1826. A second complete route was thus opened up between Birmingham and Liverpool, and Manchester, by which the distance was shortened twelve miles, and the delay occasioned by 320 feet of upward and downward lockage was done away with.

Telford was justly proud of his canals, which were the finest works of their kind that had yet been executed in England. Capacious, convenient, and substantial, they embodied his most ingenious contrivances, and his highest engineering skill. Hence we find him writing to a friend at Langholm, that, so soon as he could find "sufficient leisure from his various avocations in his own unrivalled and beloved island," it was his intention to visit France and Italy, for the purpose of ascertaining what foreigners had been able to accomplish, compared with ourselves, in the construction of canals, bridges, and harbours. "I have no doubt," said he, "as to their inferiority. During the war just brought to a close, England has not only been able to guard her own head and to carry on a gigantic struggle, but at the same time to construct canals, roads, harbours, bridges—magnificent works of peace—the like of which are probably not to be found in the world. Are not these things worthy of a nation's pride?"

CHAPTER XI.

TELFORD AS A ROAD-MAKER.

MR. TELFORD'S extensive practice as a bridge-builder led his friend Southey to designate him "Pontifex Maximus." Besides the numerous bridges erected by him in the West of England, we have found him furnishing designs for about twelve hundred in the Highlands, of various dimensions, some of stone and others of iron. His practice in bridge-building had, therefore, been of an unusually extensive character, and Southey's sobriquet was not ill applied. But besides being a great bridge-builder, Telford was also a great road-maker. With the progress of industry and trade, the easy and rapid transit of persons and goods had come to be regarded as an increasing object of public interest. Fast coaches now ran regularly between all the principal towns of England; every effort being made, by straightening and shortening the roads, cutting down hills, and carrying embankments across valleys and viaducts over rivers, to render travelling by the main routes as easy and expeditious as possible.

Attention was especially turned to the improvement of the longer routes, and to perfecting the connection of London with the chief towns of Scotland and Ireland. Telford was early called upon to advise as to the repairs of the road between Carlisle and Glasgow, which had been allowed to fall into a wretched state; as well as the formation of a new line from Carlisle, across the counties of Dumfries, Kirkcudbright, and Wigton, to Port Patrick, for the purpose of ensuring a more rapid communication with Belfast and the northern parts of Ireland. Although Glasgow had become a place of considerable wealth and importance, the roads to it, north of Carlisle, continued in a very unsatisfactory state. It was only in July, 1788, that the first mail-coach from London had driven into Glasgow by that route, when it was welcomed by a procession of the citizens on horseback, who went out several miles to meet it. But the road had been shockingly made, and before long had become almost impassable. Robert Owen states that, in 1795, it took him two days and three nights' incessant travelling to get from Manchester to Glasgow, and he mentions that the coach had to cross a well-known dangerous mountain at midnight, called Erickstane Brae, which was then always passed with fear and trembling.*

As late as the year 1814 we find a Parliamentary Committee declaring the road between Carlisle and Glasgow to be in so ruinous a state as often seriously to delay the mail and endanger the lives of travellers. The bridge over Evan Water was so much decayed, that one day the coach and horses fell through it into the river, when "one passenger was killed, the coachman survived only a few days, and several other persons were dreadfully maimed; two of the horses being also killed."† The remaining part of the bridge continued for some time unrepaired, just space enough being left for a single carriage to pass. The road trustees seemed to be helpless, and did nothing; a local subscription was tried and failed, the district passed through being very poor; but as the road was absolutely required for more than merely local purposes, it was eventually determined to undertake its reconstruction as a work of national importance, and 50,000*l.* was granted by Parliament with this object, under the provisions of the Act passed in 1816. The works were placed under Mr. Telford's charge; and an admirable road was very shortly under construction between Carlisle and Glasgow. That part of it between Hamilton and Glasgow, eleven miles in

* 'Life of Robert Owen,' by himself.

† 'Report from the Select Committee on the Carlisle and Glasgow Road,' 28th June, 1815.

length, was however left in the hands of local trustees, as was the diversion of thirteen miles at the boundary of the counties of Lanark and Dumfries, for which a previous Act had been obtained. The length of new line constructed by Mr. Telford was sixty-nine miles, and it was probably the finest piece of road which up to that time had been made.

His ordinary method of road-making in the Highlands was, first to level and drain; then, like the Romans, to lay a solid pavement of large stones, the round or broad end downwards, as close as they could be set. The points of the latter were then broken off, and a layer of stones broken to about the size of walnuts, was laid upon them, and over all a little gravel if at hand. A road thus formed soon became bound together, and for ordinary purposes was very durable.

But where the traffic, as in the case of the Carlisle and Glasgow road, was expected to be very heavy, Telford took much greater pains. Here he paid especial attention to two points: first, to lay it out as nearly as possible upon a level, so as to reduce the draught to horses dragging heavy vehicles,—one in thirty being about the severest gradient at any part of the road. The next point was to make the working, or middle portion of the road, as firm and substantial as possible, so as to bear, without shrinking, the heaviest weight likely to be brought over it. With this object he specified that the metal bed was to be formed in two layers, rising about four inches towards the centre—the bottom course being of stones (whinstone, limestone, or hard freestone), seven inches in depth. These were to be carefully set by hand, with the broadest ends downwards, all crossbonded or jointed, no stone being more than three inches wide on the top. The spaces between them were then to be filled up with smaller stones, packed by hand, so as to bring the whole to an even and firm surface. Over this a top course was to be laid, seven inches in depth, consisting of properly broken hard whinstones, none exceeding six ounces in weight, and each to be able to pass through a circular ring, two inches and a half in diameter; a binding of gravel, about an inch in thickness, being placed over all. A drain crossed under the bed of the bottom layer to the outside ditch in every hundred yards. The result was an admirably easy, firm, and dry road, capable of being travelled upon in all weathers, and standing in comparatively small need of repairs.

A similar practice was introduced in England about the same time by Mr. Macadam; and, though his method was not so thorough as that of Telford, it was usefully employed on most of the high roads throughout the kingdom. Mr. Macadam's notice was first called to the subject while acting as one of the trustees of a road in Ayrshire. Afterwards, while employed as Government agent for victualling the navy in the western parts of England, he continued the study of road-making, keeping in view the essential conditions of a compact and durable substance and a smooth surface. At that time the attention of the Legislature was not so much directed to the proper making and mending of the roads, as to suiting the vehicles to them such as they were; and they legislated backwards and forwards for nearly half a century as to the breadth of wheels. Macadam was, on the other hand, of opinion that the main point was to attend to the nature of the roads on which the vehicles were to travel. Most roads were then made with gravel, or flints tumbled upon them in their natural state, and so rounded that they had no points of contact, and rarely

J. L. MACADAM.

became consolidated. When a heavy vehicle of any sort passed over them, their loose structure presented no resistance; the material was thus completely disturbed, and they often became almost impassable. Macadam's practice was this: to break the stones into angular fragments, so that a bed several inches in depth should be formed, the material best adapted for the purpose being fragments of granite, greenstone, or basalt; to watch the repairs of the road carefully during the process of consolidation, filling up the inequalities caused by the traffic passing over it, until a hard and level surface had been obtained. Thus made, the road would last for years without further attention. In 1815 Mr. Macadam devoted himself with great enthusiasm to road-making as a profession, and being appointed surveyor-general of the Bristol roads, he had full opportunities of exemplifying his system. It proved so successful that the example set by him was quickly followed over the entire kingdom. Even the streets of many large towns were *Macadamised.* In carrying out his improvements, however, Mr. Macadam spent several thousand pounds of his own money, and in 1825, having proved this expenditure before a Committee of the House of Commons, the amount was reimbursed to him, together with an honorary tribute of two thousand pounds. Mr. Macadam died poor, but, as he himself said, "at least an honest man." By his indefatigable exertions and his success as a road-maker, by greatly saving animal labour, facilitating commercial intercourse, and rendering travelling easy and expeditious, he entitled himself to the reputation of a public benefactor.

Owing to the mountainous nature of the country through which Telford's Carlisle and Glasgow road passes, the bridges are unusually numerous and of large dimensions. Thus, the Fiddler's Burn Bridge is of three arches, one of 150 and two of 105 feet span each. There are fourteen other bridges, presenting from one to three arches, of from 20 to 90 feet span. But the most picturesque and remarkable bridge constructed by Telford in that district was upon another line of road subsequently carried out by him, in the upper part of the county of Lanark, and crossing the main line of the Carlisle and Glasgow road almost at right angles. Its northern and eastern part formed a direct line of communication between the great cattle markets of Falkirk, Crief, and Doune, and Carlisle and the West of England. It was carried over deep ravines by several lofty bridges, the most formidable of which was that across the Mouse Water at Cartland Crags, about a mile to the west of Lanark. The stream here flows through a deep rocky chasm, the sides of which are in some places about four hundred feet high. At a point where the height of the rocks is considerably less, but still most formidable, Telford spanned the ravine with the beautiful bridge represented in the engraving facing this page, its parapet being 129 feet above the surface of the water beneath.

The reconstruction of the western road from Carlisle to Glasgow, which Telford had thus satisfactorily carried out, shortly led to similar demands from the population on the eastern side of the kingdom. The spirit of road reform was now fairly on foot. Fast coaches and wheel-carriages of all kinds had become greatly improved, so that the usual rate of travelling had advanced from five or six to nine or ten miles an hour. The desire for the rapid communication of political and commercial intelligence was found to increase with the facilities for supplying it; and, urged by the public wants, the Post-Office authorities were stimulated to unusual efforts in this direction. Numerous surveys were made and roads laid out, so as to improve the main line of communication between London and Edinburgh and the intermediate towns. The first part of this road taken in hand was the worst—that lying to the north of Catterick Bridge, in Yorkshire. A new line was surveyed by West Auckland to Hexham, passing over Carter Fell to Jedburgh, and thence to Edinburgh; but was rejected as too crooked and uneven. Another was tried by Aldstone Moor and Bewcastle, and rejected for the same reason. The

CARTLAND CRAGS BRIDGE.

By R. P. LEITCH.

Page 250.

third line proposed was eventually adopted as the best, passing from Morpeth, by Wooler and Coldstream, to Edinburgh; saving rather more than fourteen miles between the two points, and securing a line of road of much more favourable gradients.

The principal bridge on this new highway was at Pathhead, over the Tyne, about eleven miles south of Edinburgh. To maintain the level, so as to avoid the winding of the road down a steep descent on one side of the valley and up an equally steep ascent on the other, Telford ran out a lofty embankment from both sides, connecting their ends by means of a spacious bridge. The structure at Pathhead is of five arches, each 50 feet span, with 25 feet rise from their springing, 49 feet above the bed of the river. Bridges of a similar character were also thrown over the deep ravines of Cranston Dean and Cotty Burn, in the same neighbourhood. At the same time a useful bridge was built on the same line of road at Morpeth, in Northumberland, over the river Wansbeck. It consisted of three arches, of which the centre one was 50 feet span, and two side-arches 40 feet each; the breadth between the parapets being 30 feet.

The advantages derived from the construction of these new roads were found to be so great, that it was proposed to do the like for the remainder of the line between London and Edinburgh; and at the instance of the Post-Office authorities, with the sanction of the Treasury, Mr. Telford proceeded to make detailed surveys of an entire new post-road between London and Morpeth. In laying it out, the main points which he endeavoured to secure were directness and flatness; and 100 miles of the proposed new Great North Road, south of York, were laid out in a perfectly straight line. This survey, which was begun in 1824, extended over several years; and all the requisite arrangements had been made for beginning the works, when the result of the locomotive competition at Rainhill, in 1829, had the effect of directing attention to that new method of travelling, fortunately in time to prevent what would have proved, for the most part, an unnecessary expenditure, on works soon to be superseded by a totally different order of things.

The most important road-improvements actually carried out under Mr. Telford's immediate superintendence were those on the western side of the island, with the object of shortening the distance and facilitating the communication between London and Dublin by way of Holyhead, as well as between London and Liverpool. At the time of the Union, the mode of transit between the capital of Ireland and the metropolis of the United Kingdom was tedious, difficult, and full of peril. In crossing the Irish Sea to Liverpool, the packets were frequently tossed about for days together. On the Irish side, there was scarcely the pretence of a port, the landing-place being within the bar of the river Liffey, inconvenient at all times, and in rough weather extremely dangerous. To avoid the long voyage to Liverpool, the passage began to be made from Dublin to Holyhead, the nearest point of the Welsh coast. Arrived there, the passengers were landed upon rugged, unprotected rocks, without a pier or landing convenience of any kind.* But the traveller's perils were not at an end,—comparatively speaking they had only begun. From Holyhead, across the island of Anglesea, there was no made road, but only a miserable track, circuitous and craggy, full of terrible jolts, round bogs and over rocks, for a distance of twenty-four miles. Having reached the Menai Strait, the passengers had again to take to an open

* A diary is preserved of a journey to Dublin from Grosvenor Square, London, 12th June, 1787, in a coach and four, accompanied by a post-chaise and pair, and five outriders. The party reached Holyhead in four days, at a cost of 75*l.* 11*s.* 3*d.* The state of intercourse between this country and the sister island at this part of the account is strikingly set forth in the following entries:—"Ferry at Bangor, 1*l.* 10*s.*; expenses of the yacht hired to carry the party across the channel, 28*l.* 7*s.* 9*d.*; duty on the coach, 7*l.* 13*s.* 4*d.*; boats on shore, 1*l.* 1*s.*; total, 114*l.* 3*s.* 4*d.*"—Roberts's 'Social History of the Southern Counties,' p. 504.

ferry-boat before they could gain the main land. The tide ran with great rapidity through the Strait, and, when the wind blew strong, the boat was liable to be driven far up or down the channel, and was sometimes swamped altogether. The perils of the Welsh roads had next to be encountered, and these were in as bad a condition at the beginning of the present century as those of the Highlands above described. Through North Wales they were rough, narrow, steep, and unprotected, mostly unfenced, and in winter almost impassable. The whole traffic on the road between Shrewsbury and Bangor was conveyed by a small cart, which passed between the two places once a week in summer. As an illustration of the state of the roads in South Wales, which were quite as bad as those in the North, we may state that, in 1803, when the late Lord Sudeley took home his bride from the neighbourhood of Welshpool to his residence only thirteen miles distant, the carriage in which the newly married pair rode stuck in a quagmire, and the occupants, having extricated themselves from their perilous situation, performed the rest of their journey on foot.

The first step taken was to improve the landing-places on both the Irish and Welsh sides of St. George's Channel, and for this purpose Mr. Rennie was employed in 1801. The result was, that Howth on the one coast, and Holyhead on the other, were fixed upon as the most eligible sites for packet stations. Improvements, however, proceeded slowly, and it was not until 1810 that a sum of 10,000*l*. was granted by Parliament to enable the necessary works to be begun. Attention was then turned to the state of the roads, and here Mr. Telford's services were called into requisition. As early as 1808 it had been determined by the Post-Office authorities to put on a mail-coach between Shrewsbury and Holyhead; but it was pointed out that the roads in North Wales were so rough and dangerous that it was doubtful whether the service could be conducted with safety. Attempts were made to enforce the law with reference to their repair, and no less than twenty-one townships were indicted by the Postmaster-General. The route was found too perilous even for a riding post, the legs of three horses having been broken in one week.* The road across Anglesea was quite as bad. Sir Henry Parnell mentioned, in 1819, that the coach had been overturned beyond Gwynder, going down one of the hills, when a friend of his was thrown a considerable distance from the roof into a pool of water. Near the post-office of Gwynder, the coachman had been thrown from his seat by a violent jolt, and broken his leg. The post-coach, and also the mail, had been overturned at the bottom of Penmyndd Hill; and the route was so dangerous that the London coachmen, who had been brought down to "work" the country, refused to continue the duty because of its excessive dangers. Of course, anything like a regular mail-service through such a district was altogether impracticable.

The indictments of the townships proved of no use; the localities were too poor to provide the means required to construct a line of road sufficient for the conveyance of mails and passengers between England and Ireland. The work was really a national one, to be carried out at the national cost. How was this best to be done? Telford recommended that the old road between Shrewsbury and Holyhead (109 miles long) should be shortened by about four miles, and made as nearly as possible on a level; the new line proceeding from Shrewsbury by Llangollen, Corwen, Bettws-y-Coed, Capel-Curig, and Bangor, to Holyhead. Mr. Telford also proposed to cross the Menai Strait by means of a cast iron bridge, hereafter to be described.

Although a complete survey was made in 1811, nothing was done for several years. The mail-coaches continued to be overturned, and stage-coaches, in the tourist season, to break down as before.† The Irish mail-coach took forty-

* 'Second Report from Committee on Holyhead Roads and Harbours,' 1810. (Parliamentary paper.)

† "Many parts of the road are extremely dangerous for a coach to

one hours to reach Holyhead from the time of its setting out from St. Martin's-le-Grand; the journey was performed at the rate of only 6¾ miles an hour, the mail arriving in Dublin on the third day. The Irish members made many complaints of the delay and dangers to which they were exposed in travelling up to town. But, although there was much discussion, no money was voted until the year 1815, when Sir Henry Parnell vigorously took the question in hand and successfully carried it through. A Board of Parliamentary Commissioners was appointed, of which he was chairman, and, under their direction, the new Shrewsbury and Holyhead road was at length commenced and carried to completion, the works extending over a period of about fifteen years. The same Commissioners exercised an authority over the roads between London and Shrewsbury; and numerous improvements were also made in the main line at various points, with the object of facilitating communication between London and Liverpool as well as between London and Dublin.

The rugged nature of the country through which the new road passed, along the slopes of rocky precipices and across inlets of the sea, rendered it necessary to build many bridges, to form many embankments, and cut away long

travel upon. At several places between Bangor and Capel-Curig there are a number of dangerous precipices without fences, exclusive of various hills that want taking down. At Ogwen Pool there is a very dangerous place where the water runs over the road, extremely difficult to pass at flooded times. Then there is Dinas Hill, that needs a side fence against a deep precipice. The width of the road is not above twelve feet in the steepest part of the hill, and two carriages cannot pass without the greatest danger. Between this hill and Rhyddlanfair there are a number of dangerous precipices, steep hills, and difficult narrow turnings. From Corwen to Llangollen the road is very narrow, long, and steep; has no side fence, except about a foot and a half of mould or dirt, which is thrown up to prevent carriages falling down three or four hundred feet into the river Dee. Stage-coaches have been frequently overturned and broken down from the badness of the road, and the mails have been overturned; but I wonder that more and worse accidents have not happened, the roads are so bad."—Evidence of Mr. William Akers, of the Post-Office, before Committee of the House of Commons, 1st June, 1815.

stretches of rock, in order to secure an easy and commodious route. The line of the valley of the Dee, to the west of Llangollen, was selected, the road proceeding along the scarped sides of the mountains, crossing from point to point by lofty embankments where necessary; and, taking into account the character of the country, it must be acknowledged that a wonderfully level road was secured. While the gradients on the old road had in some cases been as steep as 1 in 6½, passing along the edge of unprotected precipices, the new one was so laid out as to be no more than 1 in 20 at any part, while it was wide and well protected along its whole extent. Mr. Telford pursued the same system that he had adopted in the formation of the Carlisle and Glasgow road, as regards metalling, cross-draining, and fence-walling; for the latter purpose using schistus, or slate rubble-work, instead of sandstone. The largest bridges were of iron; that at Bettws-y-Coed, over the Conway—called the Waterloo Bridge, constructed in 1815—being a very fine specimen of Telford's iron bridge-work.

Those parts of the road which had been the most dangerous were taken in hand first, and, by the year 1819, the route had been rendered comparatively commodious and safe. Angles were cut off, the sides of hills were blasted away, and several heavy embankments run out across formidable arms of the sea. Thus, at Stanley Sands, near Holyhead, an embankment was formed 1300 yards long and 16 feet high, with a width of 34 feet at the top, along which the road was laid. Its breadth at the base was 114 feet, and both sides were coated with rubble stones, as a protection against storms. By the adoption of this expedient, a mile and a half was saved in a distance of six miles. Heavy embankments were also run out, where bridges were thrown across chasms and ravines, to maintain the general level. From Ty-Gwynn to Lake Ogwen, the road along the face of the rugged hill and across the river Ogwen was entirely new made, of a uniform width of 28 feet between the parapets, with an inclination of only 1 in 22 in the steepest place. A bridge was thrown over the deep chasm forming the channel of the Ogwen, the embankment being carried forward from the rock cutting, protected by high breastworks. From Capel-Curig to near the great waterfall over the river Lugwy, about a mile of new road was cut; and a still greater length from Bettws across the river Conway and along the face of Dinas Hill to Rhyddlanfair, a distance of 3 miles; its steepest descent being 1 in 22, diminishing to 1 in 45. By this improvement, the most difficult and dangerous pass along the route through North Wales was rendered safe and commodious. Another point of almost equal difficulty occurred near Ty-Nant, through the rocky pass of Glynn Duffrws, where the road was confined between steep rocks and rugged precipices: there the way was widened and flattened by blasting, and thus reduced to

ROAD DESCENT NEAR BETTWS-Y-COED.

[By Percival Skelton.]

ROAD ABOVE NANT FFRANCON, NORTH WALES.

[By Percival Skelton, after his original Drawing.]

the general level; and so on eastward to Llangollen and Chirk, where the main Shrewsbury road to London was joined.*

* The Select Committee of the House of Commons, in reporting as to the manner in which these works were carried out, stated as follows:—"The professional execution of the new works upon this road greatly surpasses anything of the same kind in these countries. The science which has been displayed in giving the general line of the road a proper inclination through a country whose whole surface consists of a succession of rocks, bogs, ravines, rivers, and precipices, reflects the greatest credit upon the engineer who has planned them; but perhaps a still greater degree of professional skill has been shown in the construction, or rather the building, of the road itself. The great attention which Mr. Telford has devoted, to give to the surface of the road one uniform and moderately convex shape, free from the smallest inequality throughout its whole breadth; the numerous land drains, and, when necessary, shores and tunnels of substantial masonry, with which all the water arising

By means of these admirable roads the traffic of North Wales continues to be mainly carried on to this day. Although railways have superseded coach-roads in the more level districts, the hilly nature of Wales precludes their formation in that quarter to any considerable extent; and even in the event of railways being constructed, a large part of the traffic of every country must necessarily continue to pass over the old high roads. Without them even railways would be of comparatively little value; for a railway station is of use chiefly because of its easy accessibility, and thus, both for passengers and merchandise, the common roads of the country are as useful as ever they were, though the main post-roads have in a great measure ceased to be employed for the purposes for which they were originally designed.

The excellence of the roads constructed by Mr. Telford through the formerly inaccessible counties of North Wales was the theme of general praise; and their superiority, compared with those of the richer and more level districts in the midland and western English counties, becoming the subject of public comment, he was called upon to execute like improvements upon that part of the post-road which extended between Shrewsbury and the metropolis. A careful survey was made of the several routes from London northward by Shrewsbury as far as Liverpool; and the short line by Coventry, being 153 miles from London to Shrewsbury, was selected as the one to be improved to the utmost.

Down to 1819, the road between London and Coventry was in a very bad state, being so laid as to become a heavy slough in wet weather. There were many steep hills which required to be cut down, in some parts of deep clay, in

from springs or falling in rain is instantly carried off; the great care with which a sufficient foundation is established for the road, and the quality, solidity, and disposition of the materials that are put upon it, are matters quite new in the system of road making in these countries."—'Report from the Select Committee on the Road from London to Holyhead in the year 1819.'

others of deep sand. A mail-coach had been tried to Banbury; but the road below Aylesbury was so bad, that the Post-Office authorities were obliged to give it up. The twelve miles from Towcester to Daventry were still worse. The line of way was covered with banks of dirt; in winter it was a puddle of from four to six inches deep—quite as bad as it had been in Arthur Young's time; and when horses passed along the road, they came out of it a mass of mud and mire.* There were also several steep and dangerous hills to be crossed; and the loss of horses by fatigue in travelling by that route at the time was very great.

Even the roads in the immediate neighbourhood of the metropolis were little better, those under the Highgate and Hampstead trust being pronounced in a wretched state. They were badly formed, on a clay bottom, and being undrained, were almost always wet and sloppy. The gravel was usually tumbled on and spread unbroken, so that the materials, instead of becoming consolidated, were only rolled about by the wheels of the carriages passing over them.

Mr. Telford applied the same methods in the reconstruction of these roads that he had already adopted in Scotland and Wales, and the same improvement was shortly felt in the more easy passage over them of vehicles of all sorts, and in the great acceleration of the mail service. At the same time, the line along the coast from Bangor, by Conway, Abergele, St. Asaph, and Holywell, to Chester, was greatly improved. As forming the mail road from Dublin to Liverpool, it was considered of importance to render it as safe and level as possible. The principal new cuts on this line were those along the rugged skirts of the huge Penmaen-Mawr; around the base of Penmaen-Bach to the town of Conway; and between St. Asaph and Holywell, to ease the ascent of Rhyall Hill.

---

* Evidence of William Waterhouse before the Select Committee, 10th March, 1819.

But more important than all, as a means of completing the main line of communication between England and Ireland, there were the great bridges over the Conway and the Menai Straits to be constructed. The dangerous ferries at those places had still to be crossed in open boats, sometimes in the night, when the luggage and mails were exposed to great risks. Sometimes, indeed, they were wholly lost, and passengers were lost with them. It was therefore determined, after long consideration, to erect bridges over these formidable straits, and Mr. Telford was employed to execute the works,—in what manner, we propose to describe in the next chapter.

CHAPTER XII.

THE MENAI AND CONWAY BRIDGES.

So long as the dangerous Straits of Menai had to be crossed in an open ferry-boat, the communication between London and Holyhea was necessarily considered incomplete. While the roads through North Wales were so dangerous as to deter travellers between England and Ireland from using that route, the completion of the remaining link of communication across the Straits was of comparatively little importance. But when those roads had, by the application of much capital, skill, and labour, been rendered so safe and convenient that the mail and stage coaches could run over them at the rate of from eight to ten miles an hour, the

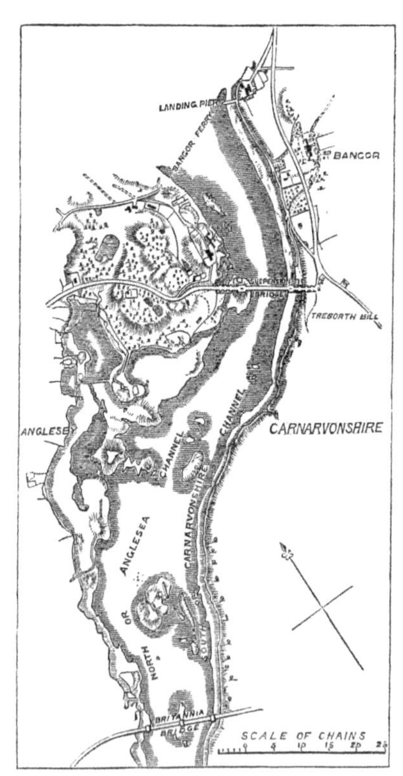

MAP OF MENAI STRAIT. [Ordnance Survey.]

bridging of the Straits became a measure of urgent public necessity. The increased traffic by this route so much increased the quantity of passengers and luggage, that the open boats were often dangerously overloaded; and serious accidents, attended with loss of life and property, came to be of frequent occurrence.

The erection of a bridge over the Straits had long been matter of speculation amongst engineers. As early as 1776, Mr. Golborne proposed his plan of an embankment with a bridge in the middle of it; and a few years later, in 1785, Mr. Nichols proposed a wooden viaduct, furnished with drawbridges at Cadnant Island. Later still, Mr. Rennie proposed his design of a cast iron bridge. But none of these plans were carried out, and the whole subject remained in abeyance until the year 1810, when a commission was appointed to inquire and report as to the state of the roads between Shrewsbury, Chester, and Holyhead. The result was, that Mr. Telford was called upon to report as to the most effectual method of bridging the Menai Strait, and thus completing the communication with the port of embarkation for Ireland.

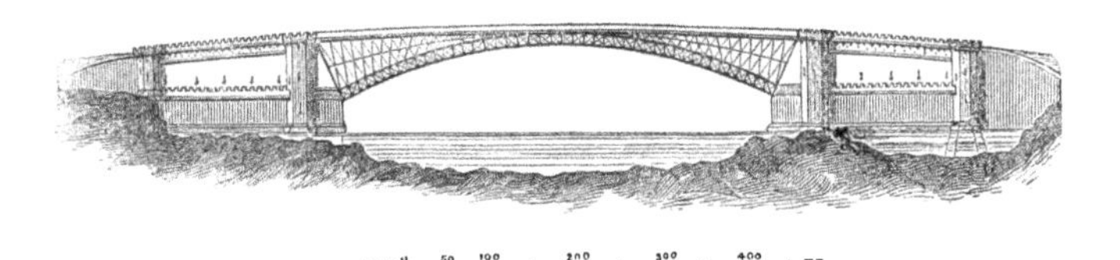

TELFORD'S PROPOSED CAST IRON BRIDGE.

Mr. Telford submitted alternative plans for a bridge over the Strait: one at the Swilly Rock, consisting of three cast iron arches of 260 feet span, with a stone arch of 100 feet span between each two iron ones, to resist their lateral thrust; and another at Ynys-y-moch, to which he himself attached the preference, consisting of a single cast iron arch of 500 feet span, the crown of the arch to be 100 feet above high water of spring tides, and the breadth of the roadway to be 40 feet.

The principal objection taken to this plan by engineers generally, was the supposed difficulty of erecting a proper centering to support the arch during construction; and the mode by which Mr. Telford proposed to overcome this may be cited in illustration of his ready ingenuity in overcoming difficulties. He proposed to suspend the centering from above instead of supporting it from below in the usual manner—a contrivance afterwards revived by another very skilful engineer, the late Mr. Brunel. Frames, 50 feet high, were to be erected on the top of the abutments, and on these, strong blocks, or rollers and chains, were to be fixed, by means of which, and by the aid of windlasses and other mechanical powers, each separate piece of centering was to be raised into, and suspended in, its proper

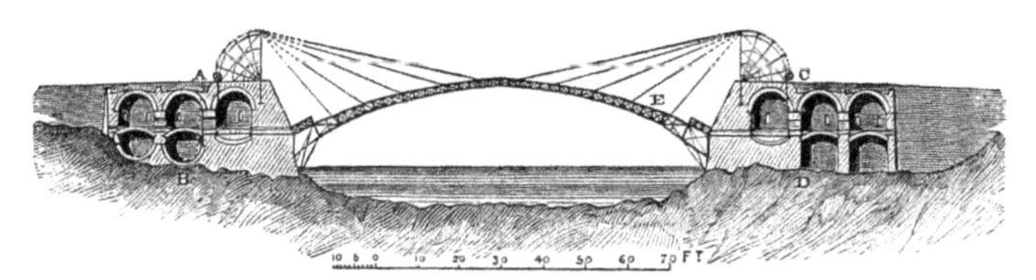

PROPOSED PLAN OF SUSPENDED CENTERING.

place. Mr. Telford regarded this method of constructing centres as applicable to stone as well as to iron arches; and indeed it is applicable, as Mr. Brunel held, to the building of the arch itself.* Mr. Telford anticipated that,

* In an article in the 'Edinburgh Review,' No. cxli., from the pen of Sir David Brewster, the writer observes:—"Mr. Telford's principle of suspending and laying down from above the centering of stone and iron bridges is, we think, a much more fertile one than even he himself supposed. With modifications, by no means considerable, and certainly practicable, it appears to us that the voussoirs or arch-stones might themselves be laid down from above, and suspended by an appropriate mechanism till the keystone was inserted. If we suppose the centering in Mr. Telford's plan to be of iron, this centering itself becomes an iron bridge, each rib of which is composed of ten pieces of fifty feet each; and by increasing the num-

if the method recommended by him were successfully adopted on the large scale proposed at Menai, all difficulties with regard to carrying bridges over deep ravines would be done away with, and a new era in bridge-building begun. For this and other reasons — but chiefly because of the much greater durability of a cast iron bridge compared with the suspension bridge afterwards adopted—it is matter of regret that he was not permitted to carry out this novel and grand design. It was, however, again objected by mariners that the bridge would seriously affect, if not destroy, the navigation of the Strait; and this plan, like Mr. Rennie's, was eventually rejected.

Several years passed, and during the interval Mr. Telford was consulted as to the construction of a bridge over Runcorn Gap on the Mersey, above Liverpool. As the river was there about 1200 feet wide, and much used for purposes of navigation, a bridge of the ordinary construction was found inapplicable. But as he was required to furnish a plan of the most suitable structure, he proceeded to consider how the difficulties of the case were to be met. The only practicable plan, he thought, was a bridge constructed on the principle of suspension. Expedients of this kind had long been employed in India and America, where wide rivers were crossed by means of bridges formed of ropes and chains; and even in this country a suspension bridge, though of a very rude kind, had long been in use near Middleton on the Tees, where, by means of two common chains stretched across the river, upon which a footway of boards

---

ber of suspending chains, these separate pieces or voussoirs having been previously joined together, either temporarily or permanently, by cement or by clamps, might be laid into their place, and kept there by a single chain till the road was completed. The voussoirs, when united, might be suspended from a general chain across the archway, and a platform could be added to facilitate the operations." This is as nearly as possible the plan afterwards revived by Mr. Brunel, and for the originality of which, we believe, he has generally the credit, though it clearly belongs to Telford.

was laid, the colliers were enabled to pass from their cottages to the colliery on the opposite bank.

Captain (afterwards Sir Samuel) Brown took out a patent for forming suspension bridges in 1817; but it appears that Telford's attention had been directed to the subject before this time, as he was first consulted respecting the Runcorn Bridge in the year 1814, when he proceeded to make an elaborate series of experiments on the tenacity of wrought iron bars, with the object of employing this material in his proposed structure. After he had made upwards of two hundred tests of malleable iron of various qualities, he proceeded to prepare his design of a bridge, which consisted of a central opening of 1000 feet span, and two side openings of 500 feet each, supported by pyramids of masonry placed near the low-water lines. The roadway was to be 30 feet wide, divided into one central footway and two distinct carriageways of 12 feet each. At the same time he prepared and submitted a model of the central opening, which satisfactorily stood the various strains which were applied to it. This Runcorn design of 1814 was of a very magnificent character, perhaps superior even to that of the Menai Suspension Bridge, afterwards erected; but unhappily the means were not forthcoming to carry it into effect. The publication of his plan and report had, however, the effect of directing public attention to the construction of bridges on the suspension principle; and many were shortly after designed and erected by Telford and other engineers in different parts of the kingdom.

Mr. Telford continued to be consulted by the Commissioners of the Holyhead Roads as to the completion of the last and most important link in the line of communication between London and Holyhead, by bridging the Straits of Menai; and at one of their meetings in 1815, shortly after the publication of his Runcorn design, the inquiry was made whether a bridge upon the same principle was not applicable in this particular case. The engineer was instructed again to examine the Straits and submit a suitable plan and estimate, which he proceeded to do in the early part of 1818. The site selected by him as the most favourable was that which had been previously fixed upon for the projected cast iron bridge, namely at Ynys-y-moch—the shores there being bold and rocky, affording easy access and excellent foundations, while by spanning the entire channel between the low-water lines, and the roadway being kept uniformly 100 feet above the highest water at spring tide, the whole of the navigable waterway would be left entirely uninterrupted. The distance between the centres of the supporting pyramids was proposed to be of the then unprecedented width of 550 feet, and the height of the pyramids 53 feet above the level of the roadway. The main chains were to be sixteen in number, with a

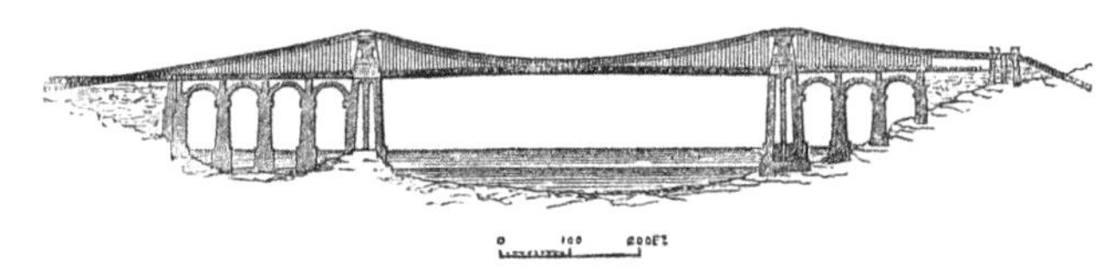

OUTLINE OF MENAI BRIDGE.

deflection of 37 feet, each composed of thirty-six bars of half-inch-square iron, so placed as to give a square of six on each side, making the whole chain about four inches in diameter, welded together for their whole length, secured by bucklings, and braced round with iron wire; while the ends of these great chains were to be secured by a mass of masonry built over stone arches between each end of the supporting piers and the adjoining shore. Four of the arches were to be on the Anglesea, and three on the Caernarvonshire side, each of them of 52 feet 6 inches span. The roadway was to be divided, as in the Runcorn design—with a carriageway 12 feet wide on each side, and a footpath of 4 feet in the middle. Mr. Telford's plan was supported by Mr. Rennie and other engineers of eminence; and the Select Committee of the House of Commons, being satisfied as to its practicability, recommended Parliament to pass a Bill and to make a grant of money to enable the work to be carried into effect.

The necessary Act passed in the session of 1819, and Mr. Telford immediately proceeded to Bangor to make preparations for beginning the works. The first proceeding was to blast off the inequalities of the surface of the rock called Ynys-y-moch, situated on the western or Holyhead side of the Strait, at that time accessible only at low water. The object was to form an even surface upon it for the foundation of the west main pier. It used to be at this point, where the Strait was narrowest, that horned cattle were driven down, preparatory to swimming them across the channel to the Caernarvon side, when the tide was weak and at its lowest ebb. The cattle were, nevertheless, often carried away, the current being too strong for the animals to contend against it.

At the same time, a landing-quay was erected on Ynys-y-moch, which was connected with the shore by an embankment carrying lines of railway. Along these, horses drew the sledges laden with stone required for the work; the material being brought in barges from the quarries opened at Penmon Point, on the north-eastern extremity of the Isle of Anglesea, a little to the westward of the northern opening of the Strait. When the surface of the rock had been levelled and the causeway completed, the first stone of the main pier was laid by Mr. W. A. Provis, the resident engineer, on the 10th of August, 1819; but not the slightest ceremony was observed on the occasion.

Later in the autumn, preparations were made for proceeding with the foundations of the eastern main pier on the Bangor side of the Strait. After excavating the beach to a depth of 7 feet, a solid mass of rock was reached, which served the purpose of an immoveable foundation for the pier. At the same, time workshops were erected; builders, artisans, and labourers were brought together from distant quarters; vessels and barges were purchased or built for the special purpose of the work; a quay was constructed at Penmon Point for loading the stones for the piers; and all the requisite preliminary arrangements were made for proceeding with the building operations in the ensuing spring.

A careful specification of the masonry work was drawn up, and the contract was let to Messrs. Stapleton and Hall; but as they did not proceed satisfactorily, and desired to be released from the contract, it was relet on the same terms to Mr. John Wilson, one of Mr. Telford's principal contractors for mason work on the Caledonian Canal. The building operations were begun with great vigour early in 1820. The three arches on the Caernarvonshire side and the four on the Anglesea side were first proceeded with. They are of immense magnitude, and occupied four years in construction, having been finished late in the autumn of 1824. These piers are 65 feet in height from high-water line to the springing of the arches, the span of each being 52 feet 6 inches. The work of the main piers also made satisfactory progress, and the masonry proceeded so rapidly that stones could scarcely be got from the quarries in sufficient quantity to keep the builders at work. By the end of June about three hundred men were employed.

The two principal piers, each 153 feet in height, upon which the main chains of the bridge were to be suspended, were built with great care and under rigorous inspection. In these, as indeed in most of the masonry of the bridge, Mr. Telford adopted the same practice which he had employed in his previous bridge structures, that of leaving large void spaces, commencing above high water mark and continuing them up perpendicularly nearly to the level of the roadway. "I have elsewhere expressed my conviction," he says, when referring to the mode of constructing these piers, "that one of the most important improvements which I have been able to introduce into masonry consists in the preference of cross-walls to rubble, in the structure of a pier, or any other edifice requiring strength. Every stone and joint in such walls is open to inspection in the progress of the work, and even afterwards, if necessary; but a solid filling of rubble conceals itself, and may be little better than a heap of rubbish confined by side walls." The walls of these main piers were built from within as well as from without all the way up, and the inside was as carefully and closely cemented with mortar as the external face. Thus the whole pier was bound firmly together, and the utmost strength given, while the weight of the superstructure upon the lower parts of the work was reduced to its minimum.

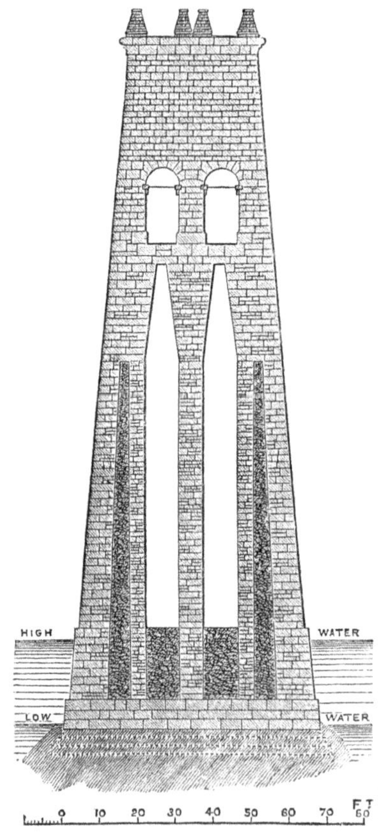

SECTION OF MAIN PIER.

Over the main piers, the small arches intended for the roadways were constructed, each being 15 feet to the springing of the arch, and 9 feet wide. Upon these arches the masonry was carried upwards, in a tapering form, to a height of 53 feet above the level of the road. As these piers were to carry the immense weight of the suspension chains, great pains were taken with their construction, and all the stones, from top to bottom, were firmly bound together with iron dowels to prevent the possibility of their being separated or bulged by the immense pressure they had to withstand.

The most important point in the execution of the details of the bridge, where the engineer had no past experience to guide him, was in the designing and fixing of the wrought iron work. Mr. Telford had continued his experiments as to the tenacity of bar iron, until he had obtained several hundred distinct tests; and at length, after the most mature deliberation, the patterns and dimensions were finally arranged by him, and the contract for the manufacture of the whole was let to Mr. Hazeldean, of Shrewsbury, in the year 1820. The iron was to be of the best Shropshire, drawn at Upton forge, and finished and proved at the works, under the inspection of a person appointed by the engineer.

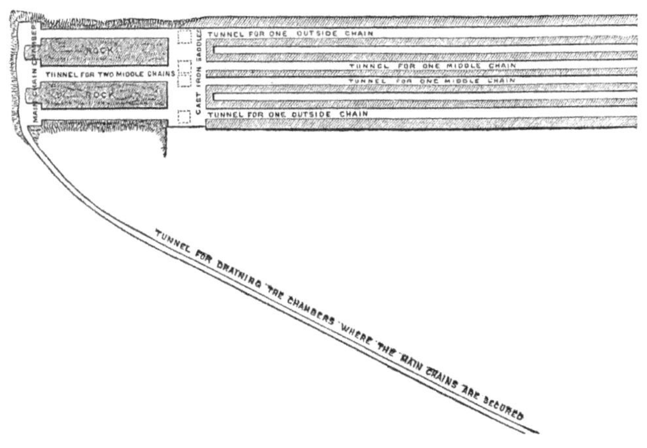

CUT SHOWING FIXING OF THE CHAINS IN THE ROCK.

The mode by which the land ends of these enormous suspension chains were rooted to the solid ground on either side of the Strait, was remarkably ingenious and effective. Three oblique tunnels were made by blasting the rock on the Anglesea side; they were each about six feet in diameter, the excavations being carried down an inclined plane to the depth of about twenty yards. A considerable width of rock lay between each tunnel, but at the bottom they were all united by a connecting horizontal avenue or cavern, sufficiently capacious to enable the workmen to fix the strong iron frames, composed principally of thick flat cast iron plates, which were engrafted deeply into the rock, and strongly bound together by the iron work passing along the horizontal avenue; so that, if the iron held, the chains could only yield by tearing up the whole mass of solid rock under which they were thus firmly bound.

A similar method of anchoring the main chains was adopted on the Caernarvonshire side. A thick bank of earth had there to be cut through, and a solid mass of masonry built in its place, the rock being situated at a greater distance from the main pier; involving a greater length of suspending chain, and a disproportion in the catenary or chord line on that side of the bridge. The excavation and masonry thereby rendered necessary proved a work of vast labour, and its execution occupied a considerable time; but by the beginning of the year 1825 the suspension pyramids, the land piers and arches, and the rock tunnels, had all been completed, and the main chains were firmly secured in them; the work being sufficiently advanced to enable the suspending of the chains to be proceeded with. This was by far the most difficult and anxious part of the undertaking.

With the same careful forethought and provision for every contingency which had distinguished the engineer's procedure in the course of the work, he had made frequent experiments to ascertain the actual power which would be required to raise the main chains to their proper curvature. A valley lay convenient for the purpose, a little to the west of the bridge on the Anglesea side. Fifty-seven of the intended vertical suspending rods, each nearly ten feet long and an inch square, having been fastened together, a piece of chain was attached to one end to make the chord line 570 feet in length; and experiments having been made and comparisons drawn, Mr. Telford ascertained that the absolute weight of one of the main chains of the bridge between the points of suspension was 23½ tons, requiring a strain of 39½ tons to raise it to its proper curvature. On this calculation the necessary apparatus required for the hoisting was prepared. The mode of action finally determined on for lifting the main chains, and fixing them into their places, was to build the central portion of each upon a raft 450 feet long and 6 feet wide, then to float it to the site of the bridge, and lift it into its place by capstans and proper tackle.

At length all was ready for hoisting the first great chain, and about the middle of April, 1825, Mr. Telford left London for Bangor to superintend the operations. An immense assemblage collected to witness the sight; greater in number than any that had been collected in the same place since the men of Anglesea, in their war-paint, rushing down to the beach, had shrieked defiance across the Straits at their Roman invaders on the Caernarvon shore. Numerous boats arrayed in gay colours glided along the waters; the day—the 26th of April—being bright, calm, and in every way propitious.

At half-past two, about an hour before high water, the raft bearing the main chain was cast off from near Treborth Mill, on the Caernarvon side. Towed by four boats, it began gradually to move from the shore, and with the assistance of the tide, which caught it at its further end, it swung slowly and majestically round to its position between the main piers, where it was moored. One end of the chain was then bolted to that which hung down the face of the Caernarvon pier; whilst the other was attached to ropes connected with strong capstans fixed on the Anglesea side, the ropes passing by means of blocks over the top of the pyramid of the Anglesea pier. The capstans for hauling in the ropes bearing the main chain, were two in number, manned by about 150 labourers. When all was ready, the signal was given to "Go along!" A band of fifers struck up a lively tune; the capstans were instantly in motion, and the men stepped round in a steady trot. All went well. The ropes gradually coiled in. As the strain increased, the pace slackened a little; but "Heave away, now she comes!" was sung out. Round went the men, and steadily and safely rose the ponderous chain.

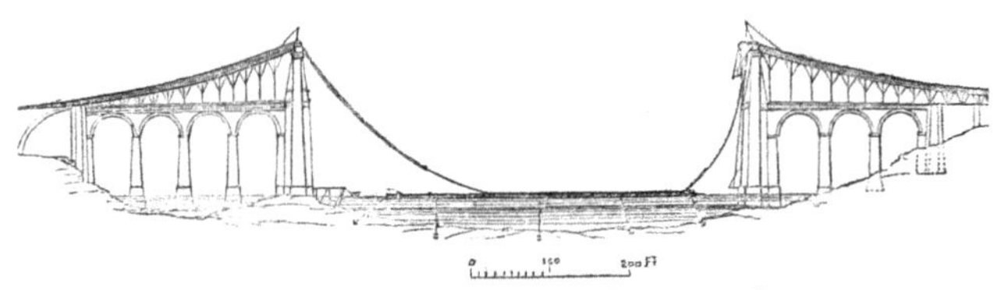

CUT OF BRIDGE, SHOWING STATE OF SUSPENSION CHAIN.

The tide had by this time turned, and bearing upon the side of the raft, now getting freer of its load, the current floated it away from under the middle of the chain still resting on it, and it swung easily off into the water. Until this moment a breathless silence pervaded the watching multitude; and nothing was heard among the working party on the Anglesea side but the steady tramp of the men at the capstans, the shrill music of the fife, and the occasional order to "Hold on!" or "Go along!" But no sooner was the raft seen floating away, and the great chain safely swinging in the air, than a tremendous cheer burst forth along both sides of the Straits.

The rest of the work was only a matter of time. The most anxious moment had passed. In an hour and thirty-five minutes after the commencement of the hoisting, the chain was raised to its proper curvature, and fastened to the land portion of it which had been previously placed over the top of the Anglesea pyramid. Mr. Telford ascended to the point of fastening, and satisfied himself that a continuous and safe connection had been formed from the Caernarvon fastening on the rock to that on Anglesea.

The announcement of the fact was followed by loud and

prolonged cheering from the workmen, echoed by the spectators, and extending along the Straits on both sides, until it seemed to die away along the shores in the distance. Three foolhardy workmen, excited by the day's proceedings, had the temerity to scramble along the upper surface of the chain—which was only nine inches wide and formed a curvature of 590 feet—from one side of the Strait to the other! *

Far different were the feelings of the engineer who had planned this magnificent work. Its failure had been predicted; and, like Brindley's Barton Viaduct, it had been freely spoken of as a "castle in the air." Telford had, it is true, most carefully tested every part by repeated experiment, and so conclusively proved the sufficiency of the iron chains to bear the immense weight they would have to support, that he was thoroughly convinced as to the soundness of his principles of construction, and satisfied that, if rightly manufactured and properly put together, the chains would hold, and that the piers would sustain them. Still there was necessarily an element of uncertainty in the undertaking. It was the largest structure of the kind that had ever been attempted. There was the contingency of a flaw in the iron; some possible scamping in the manufacture; some little point which, in the multiplicity of details to be attended to, he might have overlooked, or which his subordinates might have neglected. It was, indeed, impossible but that he should feel intensely anxious as to the result of the day's operations. Mr. Telford afterwards stated to a friend, only a few months before his death, that

* A correspondent informs us of a still more foolhardy exploit performed on the occasion. He says, "Having been present, as a boy from Bangor grammar school, on the 26th of April, when the first chain was carried across, an incident occurred which made no small impression on my mind at the time. After the chain had reached its position, a cobbler of the neighbourhood crawled to the centre of the curve, and there finished a pair of shoes; when, having completed his task, he returned in safety to the Caernarvon side! I need not say that we schoolboys appreciated his feat of foolhardiness far more than Telford's master work."

for some time previous to the opening of the bridge, his anxiety was so great that he could scarcely sleep; and that a continuance of that condition must have very soon completely undermined his health. We are not, therefore, surprised to learn that when his friends rushed to congratulate him on the result of the first day's experiment, which decisively proved the strength and solidity of the

MENAI BRIDGE.
[By Percival Skelton, after his original Drawing.]

bridge, they should have found the engineer on his knees engaged in prayer. A vast load had been taken off his mind; the perilous enterprise of the day had been accomplished without loss of life; and his spontaneous act was thankfulness and gratitude.

The suspension of the remaining fifteen chains was accomplished without difficulty. The last was raised and fixed on the 9th of July, 1825, when the entire line was completed. On fixing the final bolt, a band of music descended from the top of the suspension pier on the Anglesea side to a scaffolding erected over the centre of the curved part of the chains, and played the National Anthem amidst the cheering of many thousand persons assembled along the shores of the Strait: while the workmen marched in procession along the bridge, on which a temporary platform had been laid, and the *St. David* steam-packet of Chester passed under the chains towards the Smithy Rocks and back again, thus re-opening the navigation of the Strait.

In August the road platform was commenced, and in September the trussed bearing bars were all suspended. The road was constructed of timber in a substantial manner, the planking being spiked together, with layers of patent felt between the planks, and the carriage way being protected by oak guards placed seven feet and a half apart. Side railings were added; the toll-houses and approach-roads were completed by the end of the year; and the bridge was opened for public traffic on Monday, the 30th of January, 1826, when the London and Holyhead mail-coach passed over it for the first time, followed by the Commissioners of the Holyhead roads, the engineer, several stage-coaches, and a multitude of private persons too numerous to mention.

We may briefly add a few facts as to the quantities of materials used, and the dimensions of this remarkable structure. The total weight of iron was 2187 tons, in 33,265 pieces. The total length of the bridge is 1710 feet, or nearly a third of a mile; the distance between the points of suspension of the main bridge being 579 feet. The total sum expended by Government in its erection, including the embankment and about half a mile of new line of road on the Caernarvon side, together with the toll-houses, was 120,000*l.*

Notwithstanding the wonders of the Britannia Bridge subsequently erected by Robert Stephenson for the passage across the same strait of the Chester and Holyhead Railway, the Menai Bridge of Telford is by far the most picturesque object. "Seen as I approached it," says Mr. Roscoe, "in the clear light of an autumnal sunset, which threw an autumnal splendour on the wide range of hills beyond, and the sweep of richly variegated groves and plantations which covered their base—the bright sun, the rocky picturesque foreground, villas, spires, and towers here and there enlivening the prospect—the Menai Bridge appeared more like the work of some great magician than the mere result of man's skill and industry."

CONWAY SUSPENSION BRIDGE. [By Percival Skelton.]

Shortly after the Menai Bridge was begun, it was determined by the Commissioners of the Holyhead road that a bridge of similar design should be built over the estuary of the Conway, immediately opposite the old castle at that place, and which had formerly been crossed by an open ferry boat. The first stone was laid on the 3rd of April, 1822, and the works having proceeded satisfactorily, the bridge and embankment approaching it were completed by the summer of 1826. But the operations being of the same kind as those connected with the larger structure above described, though of a much less difficult character, it is unnecessary to enter into any details as to the several stages of its construction. In this bridge the width between the centres of the supporting towers is 327 feet, and the height of the under side of the roadway above high water of spring tides only 15 feet. The heaviest work was an embankment as its eastern approach, 2015 feet in length and about 300 feet in width at its highest part.

It will be seen, from the view of the bridge given on the opposite page, that it is a highly picturesque structure, and combines, with the estuary which it crosses, and the ancient castle of Conway, in forming a landscape that is rarely equalled.

CHAPTER XIII.

Docks, Drainage, and Bridges.

It will have been observed, from the preceding narrative, how much had already been accomplished by skill and industry towards opening up the material resources of the kingdom. The stages of improvement which we have recorded indeed exhibit a measure of the vital energy which has from time to time existed in the nation. In the earlier periods of engineering history, the war of man was with nature. The sea was held back by embankments. The Thames, instead of being allowed to overspread the wide marshes on either bank, was confined within limited bounds, by which the navigable depth of its channel was increased, at the same time that a wide extent of land was rendered available for agriculture.

In those early days, the great object was to render the land more habitable, comfortable, and productive. Marshes were reclaimed, and wastes subdued. But so long as the country remained comparatively closed against communication, and intercourse was restricted by the want of bridges and roads, improvement was extremely slow. For, while roads are the consequence of civilisation, they are also among its most influential causes. We have seen even the blind Metcalf acting as an effective instrument of progress in the northern counties by the formation of long lines of road. Brindley and the Duke of Bridgewater carried on the work in the same districts, and conferred upon the north and north-west of England the blessings of cheap and effective water communication. Smeaton followed and carried out similar undertakings in still remoter places, joining the east and west coasts of Scotland by the

Forth and Clyde Canal, and building bridges in the far north. Rennie made harbours, built bridges, and hewed out docks for shipping, the increase in which had kept pace with the growth of our home and foreign trade. He was followed by Telford, whose long and busy life, as we have seen, was occupied in building bridges and making roads in all directions, in districts of the country formerly inaccessible, and therefore comparatively barbarous. At length the wildest districts of the Highlands and the most rugged mountain valleys of North Wales were rendered as easy of access as the comparatively level counties in the immediate neighbourhood of the metropolis.

During all this while, the wealth and industry of the country had been advancing with rapid strides. London had grown in population and importance. Many improvements had been effected in the river, but the dock accommodation was still found insufficient; and, as the recognised head of his profession, Mr. Telford, though now grown old and fast becoming infirm, was called upon to supply the requisite plans. He had been engaged upon great works for upwards of thirty years, previous to which he had led the life of a working mason. But he had been a steady, temperate man all his life; and though nearly seventy, when consulted as to the proposed new docks, his mind was as able to deal with the subject in all its bearings as it had ever been; and he undertook the work.

In 1824 a new Company was formed to provide a dock nearer to the heart of the City than any of the existing ones. The site selected was the space between the Tower and the London Docks, which included the property of St. Katherine's Hospital. The whole extent of land available was only twenty-seven acres of a very irregular figure, so that when the quays and warehouses were laid out, it was found that only about ten acres remained for the docks; but these, from the nature of the ground, presented an unusual amount of quay room. The necessary Act was obtained in 1825; the works were begun in the following year; and on the 25th of October, 1828, the new docks were completed and opened for business.

The St. Katherine Docks communicate with the river by means of an entrance tide-lock, 180 feet long and 45 feet wide, with three pairs of gates, admitting either one very large or two small vessels at a time. The lock-entrance and the sills under the two middle lock-gates were fixed at the depth of ten feet under the level of low water of ordinary spring tides. The formation of these dock-entrances was a work of much difficulty, demanding great skill on the part of the engineer. It was necessary to excavate the ground to a great depth below low water for the purpose of getting in the foundations, and the cofferdams were therefore of great strength, to enable them, when pumped out by the steam-engine, to resist the lateral pressure of forty feet of water at high tide. The difficulty was, however, effectually overcome, and the wharf walls, locks, sills and bridges of the St. Katherine Docks are generally regarded as a master-piece of harbour construction. Alluding to the rapidity with which the works were completed, Mr. Telford says: "Seldom, indeed never within my knowledge, has there been an instance of an undertaking of this magnitude, in a very confined situation, having been perfected in so short a time; . . . . but, as a practical engineer, responsible for the success of difficult operations, I must be allowed to protest against such haste, pregnant as it was, and ever will be, with risks, which, in more instances than one, severely taxed all my experience and skill, and dangerously involved the reputation of the directors as well as of their engineer."

Among the remaining bridges executed by Mr. Telford, towards the close of his professional career, may be mentioned those of Tewkesbury and Gloucester. The former town is situated on the Severn, at its confluence with the river Avon, about eleven miles above Gloucester. The surrounding district was rich and populous; but being intersected by a large river, without a bridge, the inhabitants applied to Parliament for powers to provide so necessary a convenience. The design first proposed by a local architect was a bridge of three arches; but Mr. Telford, when called upon to advise the trustees, recommended that, in order to interrupt the navigation as little as possible, the river should be spanned by a single arch; and he submitted a design of such a character, which was approved and subsequently erected. It was finished and opened in April, 1826.

This is one of the largest as well as most graceful of Mr. Telford's numerous cast iron bridges. It has a single span of 170 feet, with a rise of only 17 feet, consisting of six ribs of about three feet three inches deep, the spandrels being filled in with light diagonal work. The narrow Gothic arches in the masonry of the abutments give the bridge a very light and graceful appearance, at the same time that they afford an enlarged passage for the high river floods.

The bridge at Gloucester consists of one large stone arch of 150 feet span. It replaced a structure of great antiquity, of eight arches, which had stood for about 600 years. The roadway over it was very narrow, and the number of piers in the river and the small dimensions of the arches offered considerable obstruction to the navigation. To give the largest amount of waterway, and at the same time reduce the gradient of the road over the bridge to the greatest extent, Mr. Telford adopted the following expedient. He made the general body of the arch an ellipse, 150 feet on the chord-line and 35 feet rise, while the voussoirs, or external archstones, being in the form of a segment, have the same chord, with only 13 feet rise. "This complex form," says Mr. Telford, "converts each side of the vault of the arch into the shape of the entrance of a pipe, to suit the contracted passage of a fluid, thus lessening the flat surface opposed to the current of the river whenever the tide or upland flood rises above the springing of the middle of the ellipse, that being at four feet above low water;

whereas the flood of 1770 rose twenty feet above low water of an ordinary spring-tide, which, when there is no upland flood, rises only eight or nine feet."* The bridge was finished and opened in 1828.

The last structures erected after our engineer's designs were at Edinburgh and Glasgow: his Dean Bridge at the former place, and his Jamaica Street Bridge at the latter, being regarded as among his most successful works. Since his employment as a journeyman mason at the building of the houses in Princes Street, Edinburgh, the New Town

DEAN BRIDGE, EDINBURGH. [By E. M. Wimperis.]

had spread in all directions. At each visit to it on his way to or from the Caledonian Canal or the northern harbours, he had been no less surprised than delighted at the architectural improvements which he found going forward. A

* 'Telford's Life,' p. 261.

new quarter had risen up during his lifetime, and had extended northward and westward in long lines of magnificent buildings of freestone, until in 1829 its further progress was checked by the deep ravine running along the back of the New Town, in the bottom of which runs the little Water of Leith. It was determined to throw a stone bridge across this stream, and Telford was called upon to supply the design. The point of crossing the valley was immediately behind Moray Place, which stands almost upon its verge, the sides being bold, rocky, and finely wooded. The situation was well adapted for a picturesque structure, such as Telford was well able to supply. The depth of the ravine to be spanned involved great height in the piers, the roadway being 106 feet above the level of the stream. The bridge was of four arches of 90 feet span each, and its total length 447 feet; the breadth between the parapets for the purposes of the roadway and footpaths being 39 feet.* It was completed and opened in December, 1831.

But the most important, as it was the last, of Mr. Telford's stone bridges was that erected across the Clyde at the Broomielaw, Glasgow. Little more than fifty years since, the banks of the river at that place were literally covered with broom — and hence its name — while the stream was scarcely deep enough to float a herring-buss. Now, the Broomielaw is a quay frequented by ships of the largest burden, and bustling with trade and commerce. Skill and enterprise have deepened the Clyde, dredged away its shoals, built quays and wharves along its banks, and rendered it one of the busiest streams in the world,

* The piers are built internally with hollow compartments, as at the Menai Bridge, the side walls being 3 feet thick and the cross walls 2 feet. Projecting from the piers and abutments are pilasters of solid masonry. The main arches have their springing 70 feet from the foundations, and rise 30 feet: and at 20 feet higher, other arches, of 96 feet span and 10 feet rise, are constructed; the face of these, projecting before the main arches and spandrels, producing a distinct external soffit of 5 feet in breadth. This, with the peculiar piers, constitutes the principal distinctive feature in the bridge.

It has become a great river thoroughfare, worked by steam. On its waters the first steamboat ever constructed for purposes of traffic in Europe was launched by Henry Bell in 1812; and the Clyde boats to this day enjoy the highest prestige.

The deepening of the river at the Broomielaw had led to a gradual undermining of the foundations of the old bridge, which was situated close to the principal landing-place. A little above it, was an ancient overfall weir, which had also contributed to scour away the foundations of the piers. Besides, the bridge was felt to be narrow, inconvenient, and ill-adapted for accommodating the immense traffic passing across the Clyde at that point. It was, therefore, determined to take down the old structure, and build a new one; and Mr. Telford was called upon to supply the design. The foundation was laid with great ceremony on the 18th of March, 1833, and the new bridge was completed and opened on the 1st of January, 1836, rather more than a year after the engineer's death. It is a very fine work, consisting of seven arches, segments of circles, the central arch being 58 feet 6 inches; the span of the adjoining arches diminishing to 57 feet 9 inches, 55 feet 6 inches, and 52 feet respectively. It is 560 feet in length, with an open waterway of 389 feet, and its total width of carriageway and footpath is 60 feet, or wider, at the time it was built, than any river bridge in the kingdom.

Like most previous engineers of eminence—like Perry, Brindley, Smeaton, and Rennie—Mr. Telford was in the course of his life extensively employed in the drainage of the Fen districts. He had been jointly concerned with Mr. Rennie in carrying out the important works of the Eau Brink Cut, and at Mr. Rennie's death he succeeded to much of his practice as consulting engineer.

It was principally in designing and carrying out the drainage of the North Level that Mr. Telford distinguished himself in Fen drainage. The North Level includes all that part of the Great Bedford Level situated between

GLASGOW BRIDGE.

By R. P LEITCH.

Page 286

Morton's Leam and the river Welland, comprising about 48,000 acres of land. The river Nene, which brings down from the interior the rainfall of almost the entire county of Northampton, flows through nearly the centre of the district. In some places the stream is confined by embankments, in others it flows along artificial cuts, until it enters the great estuary of the Wash, about five miles below Wisbeach. This town is situated on another river which flows through the Level, called the Old Nene. Below the point of junction of these rivers with the Wash, and still more to seaward, was South Holland Sluice, through which the waters of the South Holland Drain entered the estuary. At that point a great mass of silt had accumulated, which tended to choke up the mouths of the rivers further inland, rendering their navigation difficult and precarious, and seriously interrupting the drainage of the whole lowland district traversed by both the Old and New Nene. Indeed the sands were accumulating at such a rate, that the outfall of the Wisbeach River threatened to become completely destroyed.

Such being the state of things, it was determined to take the opinion of some eminent engineer, and Mr. Rennie was employed to survey the district and recommend a measure for the remedy of these great evils. He performed this service in his usually careful and masterly manner; but as the method which he proposed, complete though it was, would have seriously interfered with the trade of Wisbeach, by leaving it out of the line of navigation and drainage which he proposed to open up, the corporation of that town determined to employ another engineer; and Mr. Telford was selected to examine and report upon the whole subject, keeping in view the improvement of the river immediately adjacent to the town of Wisbeach.

Mr. Telford confirmed Mr. Rennie's views to a large extent, more especially with reference to the construction of an entirely new outfall, by making an artificial channel from Kindersley's Cut to Crab-Hole Eye anchorage, by which a level lower by nearly twelve feet would be secured for the outfall waters; but he preferred leaving the river open to the tide as high as Wisbeach, rather than place a lock with draw-doors at Lutton Leam Sluice, as had been proposed by Mr. Rennie. He also suggested that the acute angle at the Horseshoe be cut off and the river deepened up to the bridge at Wisbeach, making a new cut along the bank on the south side of the town, which should join the river again immediately above it, thereby converting the intermediate space, by draw-doors and the usual contrivances, into a floating dock. Though this plan was approved by the parties interested in the drainage, to Telford's great mortification it was opposed by the corporation of Wisbeach, and like so many other excellent schemes for the improvement of the Fen districts, it eventually fell to the ground.

The cutting of a new outfall for the river Nene, however, could not much longer be delayed without great danger to the reclaimed lands of the North Level, which, but for some relief of the kind, must shortly have become submerged and reduced to their original waste condition. The subject was revived in 1822, and Mr. Telford was again called upon, in conjunction with Sir John Rennie, whose father had died in the preceding year, to submit a plan of a new Nene Outfall; but it was not until the year 1827 that the necessary Act was obtained, and then only with great difficulty and cost, in consequence of the opposition of the town of Wisbeach. The works consisted principally of a deep cut or canal, about six miles in length, penetrating far through the sandbanks into the deep waters of the Wash. They were begun in 1828, and brought to completion in 1830, with the most satisfactory results. A greatly improved outfall was secured by thus carrying the mouths of the rivers out to sea, and the drainage of the important agricultural districts through which the Nene flows was greatly benefited; while at the same time nearly 6000 acres of valuable corn-growing land were added to the county of Lincoln.

But the opening of the Nene Outfall was only the first of a series of improvements which eventually included the whole of the valuable lands of the North Level, in the district situated between the Nene and the Welland. The opening at Gunthorpe Sluice, which was the outfall for the waters of the Holland Drain, was not less than eleven feet three inches above low water at Crab-Hole; and it was therefore obvious that by lowering this opening a vastly improved drainage of the whole of the level district, extending from twenty to thirty miles inland, for which that sluice was the artificial outlet, would immediately be secured. Urged by Mr. Telford, an Act for the purpose of carrying out the requisite improvement was obtained in 1830, and the excavations having been begun shortly after, were completed in 1834.

A new cut was made from Clow's Cross to Gunthorpe Sluice, in place of the winding course of the old Shire Drain; besides which, a bridge was erected at Cross Keys, or Sutton Wash, and an embankment was made across the Salt Marshes, forming a high road, which, with the bridges previously erected at Fossdyke and Lynn, effectually connected the counties of Norfolk and Lincoln. The result of the improved outfall was what the engineer had predicted. A thorough natural drainage was secured for an extensive district, embracing nearly a hundred thousand acres of fertile land, which had before been very ineffectually though expensively cleared of the surplus water by means of windmills and steam-engines. The productiveness of the soil was greatly increased, and the health and comfort of the inhabitants promoted to an extent that surpassed all previous expectation.

The whole of the new cuts were easily navigable, being from 140 to 200 feet wide at bottom, whereas the old outlets had been variable and were often choked with shifting sand. The district was thus effectually opened up for navigation, and a convenient transit afforded for coals and other articles of consumption. Wisbeach became accessible to vessels of much larger burden, and in the course of a few years after the construction of the Nene Outfall, the trade of the port had more than doubled. Mr. Telford himself, towards the close of his life, spoke with natural pride of the improvements which he had thus been in so great a measure instrumental in carrying out, and which had so materially promoted the comfort, prosperity, and welfare of a very extensive district.*

We may mention, as a remarkable effect of the opening of the new outfall, that in a few hours the lowering of the waters was felt throughout the whole of the Fen level. The sluggish and stagnant drains, cuts, and leams in far distant places, began actually to *flow;* and the sensation created was such, that at Thorney, near Peterborough, some fifteen miles from the sea, the intelligence penetrated even to the congregation then sitting in church—for it was Sunday morning —that "the waters were running!" when immediately the whole flocked out, parson and all, to see the great sight, and acknowledge the blessings of science. A humble Fen poet of the last century thus quaintly predicted the moral results likely to arise from the improved drainage of his native district:—

> "With a change of elements suddenly
> There shall a change of men and manners be;
> Hearts thick and tough as hides shall feel remorse,
> And souls of sedge shall understand discourse;
> New hands shall learn to work, forget to steal,
> New legs shall go to church, new knees to kneel."

* "The Nene Outfall channel," says Mr. Tycho Wing, "was projected by the late Mr. Rennie in 1814, and executed jointly by Mr. Telford and the present Sir John Rennie. But the scheme of the North Level Drainage was eminently the work of Mr. Telford, and was undertaken upon his advice and responsibility, when only a few persons engaged in the Nene Outfall believed that the latter could be made, or if made, that it could be maintained. Mr. Telford distinguished himself by his foresight and judicious counsels at the most critical periods of that great measure, by his unfailing confidence in its success, and by the boldness and sagacity which prompted him to advise the making of the North Level drainage, in full expectation of the results for the sake of which the Nene Outfall was undertaken, and which are now realised to the extent of the most sanguine hopes."

The prophecy has indeed been fulfilled. The barbarous race of Fen-men has disappeared before the skill of the engineer. As the land has been drained, the half-starved fowlers and fen-roamers have subsided into the ranks of steady industry—become farmers, traders, and labourers. The plough has passed over the bed of Holland Fen, and the agriculturist reaps his increase more than a hundred fold. Wide watery wastes, formerly abounding in fish, are now covered with waving crops of corn every summer. Sheep graze on the dry bottom of Whittlesea Mere, and kine low where not many years since the silence of the waste was only disturbed by the croaking of frogs and the screaming of wild fowl. All this has been the result of the science of the engineer, the enterprise of the landowner, and the industry of our peaceful army of skilled labourers.*

* Now that the land actually won has been made so richly productive, the engineer is at work with magnificent schemes of reclamation of lands at present submerged by the sea. The Norfolk Estuary Company have a scheme for reclaiming 50,000 acres; the Lincolnshire Estuary Company, 30,000 acres; and the Victoria Level Company, 150,000 acres—all from the estuary of the Wash. By the process called warping, the land is steadily advancing upon the ocean, and before many years have passed, thousands of acres of the Victoria Level will have been reclaimed for purposes of agriculture.

CHAPTER XIV.

SOUTHEY'S TOUR IN THE HIGHLANDS.

WHILE Telford's Highland works were in full progress, he persuaded his friend Southey, the Poet Laureate, to accompany him on one of his visits of inspection, as far north as the county of Sutherland, in the autumn of 1819. Mr. Southey, as was his custom, made careful notes of the tour, which have been preserved,* and consist in a great measure of an interesting *résumé* of the engineer's operations in harbour-making, road-making, and canal-making north of the Tweed.

Southey reached Edinburgh by tho Carlisle mail about the middle of August, and was there joined by Mr. Telford, and Mr. and Mrs. Rickman,† who were to accompany him on the journey. They first proceeded to Linlithgow, Bannockburn,‡ Stirling, Callendar, the Trosachs, and round by the head of Loch Earn to Killin, Kenmore, and by Aberfeldy to Dunkeld. At the latter place, the poet admired Telford's beautiful bridge, which forms a fine feature in the foreground of the incomparable picture which the scenery of Dunkeld always presents in whatever aspect it is viewed.

From Dunkeld the party proceeded to Dundee, along the left bank of the Firth of Tay. The works connected with

* We have been indebted to Mr. Robert Rawlinson, C.E., in whose possession the MS. now is, for the privilege of inspecting it, and making the above abstract, which we have the less hesitation in giving as it has not before appeared in print.

† Mr. Rickman was the Secretary to the Highland Roads Commission.

‡ Referring to the famous battle of Bannockburn, Southey writes—"This is the only great battle that ever was lost by the English. At Hastings there was no disgrace. Here it was an army of lions commanded by a stag."

the new harbour were in active progress, and the engineer lost no time in taking his friend to see them. Southey's account is as follows:—

"Before breakfast I went with Mr. Telford to the harbour, to look at his works, which are of great magnitude and importance: a huge floating dock, and the finest graving dock I ever saw. The town expends 70,000*l.* on these improvements, which will be completed in another year. What they take from the excavations serves to raise ground which was formerly covered by the tide, but will now be of the greatest value for wharfs, yards, &c. The local authorities originally proposed to build fifteen piers, but Telford assured them that three would be sufficient; and, in telling me this, he said the creation of fifteen new Scotch *peers* was too strong a measure. . . .

"Telford's is a happy life; everywhere making roads, building bridges, forming canals, and creating harbours—works of sure, solid, permanent utility; everywhere employing a great number of persons, selecting the most meritorious, and putting them forward in the world in his own way."

After the inspection at Dundee was over, the party proceeded on their journey northward, along the east coast:—

"Near Gourdon or Bervie harbour, which is about a mile and a half on this side the town, we met Mr. Mitchell and Mr. Gibbs, two of Mr. Telford's aides-de-camp, who had come thus far to meet him. The former he calls his 'Tartar,' from his cast of countenance, which is very much like a Tartar's, as well as from his Tartar-like mode of life; for, in his office of overseer of the roads, which are under the management of the Commissioners, he travels on horseback not less than 6000 miles a year. Mr. Telford found him in the situation of a working mason, who could scarcely read or write; but noticing him for his good conduct, his activity, and his firm steady character, he has brought him forward; and Mitchell now holds a post of respectability and importance, and performs his business with excellent ability."

After inspecting the little harbour of Bervie, one of the first works of the kind executed by Telford for the Commissioners, the party proceeded by Stonehaven, and from thence along the coast to Aberdeen. Here the harbour works were visited and admired:—

"The quay," says Southey, "is very fine; and Telford has carried out his pier 900 feet beyond the point where Smeaton's terminated. This great work, which has cost 100,000*l.*, protects the entrance of the harbour from the whole force of the North Sea. A ship was entering it at the time of our visit, the *Prince of Waterloo.* She had been to America; had discharged her cargo at London; and we now saw her reach her own port in safety—a joyous and delightful sight."

The next point reached was Banff, along the Don and the line of the Inverury Canal:—

"The approach to Banff is very fine,"* says Southey, "by the Earl of Fife's grounds, where the trees are surprisingly grown, considering how near they are to the North Sea; Duff House—a square, odd, and not unhandsome pile, built by Adams (one of the Adelphi brothers), some forty years ago; a good bridge of seven arches by Smeaton; the open sea, not as we had hitherto seen it, grey under a leaden sky, but bright and blue in the sunshine; Banff on the left of the bay; the River Doveran almost lost amid banks of shingle, where it enters the sea; a white and tolerably high shore extending eastwards; a kirk, with a high spire which serves as a sea-mark; and, on the point, about a mile to the east, the town of Macduff. At Banff, we at once went to the pier, about half finished, on which 15,000*l.* will be expended, to the great benefit of this clean, cheerful, and active little town. The pier was a busy scene; hand-carts going to and fro over the railroads, cranes at work charging and discharging, plenty of workmen, and fine masses of red granite from the Peterhead quarries. The quay was almost covered with barrels of herrings, which women were busily employed in salting and packing."

The next visit was paid to the harbour works at Cullen, which were sufficiently advanced to afford improved shelter for the fishing vessels of the little port:—

"When I stood upon the pier at low water," says Southey, "seeing the tremendous rocks with which the whole shore is bristled, and the open sea to which the place is exposed, it was with a proud feeling that I saw the first talents in the world employed by the British Government in works of such unostentatious, but great, immediate, palpable, and permanent utility. Already their excel-

* See View of Banff facing p. 216.

lent effects are felt. The fishing vessels were just coming in, having caught about 300 barrels of herrings during the night. . . .

"However the Forfeited Estates Fund may have been misapplied in past times, the remainder could not be better invested than in these great improvements. Wherever a pier is needed, if the people or the proprietors of the place will raise one-half the necessary funds, Government supplies the other half. On these terms, 20,000*l.* are expending at Peterhead, and 14,000*l.* at Frazerburgh; and the works which we visited at Bervie and Banff, and many other such along this coast, would never have been undertaken without such aid; public liberality thus inducing private persons to tax themselves heavily, and expend with a good will much larger sums than could have been drawn from them by taxation."

From Cullen, the travellers proceeded in gigs to Fochabers, thence by Craigellachie Bridge, which Southey greatly admired, along Speyside, to Ballindalloch and Inverallen, where Telford's new road was in course of construction across the moors towards Forres. The country for the greater part of the way was a wild waste, nothing but mountains and heather to be seen; yet the road was as perfectly made and maintained as if it had lain through a very Goschen. The next stages were to Nairn and Inverness, from whence they proceeded to view the important works constructed at the crossing of the River Beauly:—

"At Lovat Bridge," says Southey, "we turned aside and went four miles up the river, along the Strathglass road—one of the new works, and one of the most remarkable, because of the difficulty of constructing it, and also because of the fine scenery which it commands. . . . .

"Lovat Bridge, by which we returned, is a plain, handsome structure of five arches, two of 40 feet span, two of 50, and the centre one of 60. The curve is as little as possible. I learnt in Spain to admire straight bridges; but Mr. Telford thinks there always ought to be some curve to enable the rain water to run off, and because he would have the outline look like the segment of a large circle, resting on the abutments. A double line over the arches gives a finish to the bridge, and perhaps looks as well, or almost as well, as balustrades, for not a sixpence has been allowed for ornament on these works. The sides are protected by water-wings, which are embankments of stone, to prevent the floods from extending on either side, and attacking the flanks of the bridge."

Nine miles further north, they arrived at Dingwall, near which a bridge similar to that at Beauly, though wider, had been constructed over the Conan. From thence they proceeded to Invergordon, to Ballintraed (where another pier for fishing boats was in progress), to Tain, and thence to Bonar Bridge, over the Sheir, twenty-four miles above the entrance to the Dornoch Frith, where an iron bridge, after the same model as that of Craigellachie, had been erected. This bridge is of great importance, connecting as it does the whole of the road traffic of the northern counties with the south. Southey speaks of it as

"A work of such paramount utility that it is not possible to look at it without delight. A remarkable anecdote," he continues, "was told me concerning it. An inhabitant of Sutherland, whose father was drowned at the Mickle Ferry (some miles below the bridge) in 1809, could never bear to set foot in a ferry-boat after the catastrophe, and was consequently cut off from communication with the south until this bridge was built. He then set out on a journey. 'As I went along the road by the side of the water,' said he, 'I could see no bridge. At last I came in sight of something like a spider's web in the air. If this be it, thought I, it will never do! But, presently, I came upon it; and oh! it is the finest thing that ever was made by God or man!'"

Sixteen miles north-east of Bonar Bridge, Southey crossed Fleet Mound, another ingenious work of his friend Telford, but of an altogether different character. It was thrown across the River Fleet, at the point at which it ran into the estuary or little land-locked bay outside, known as Loch Fleet. At this point there had formerly been a ford; but as the tide ran far inland, it could only be crossed at low water, and travellers had often to wait for hours before they could proceed on their journey. The embouchure being too wide for a bridge, Telford formed an embankment across it, 990 yards in length, providing four flood-gates, each 12 feet wide, at its north end, for the egress of the inland waters. These gates opened outwards, and they were so hung as to shut with the rising of the tide. The holding back of the sea from the land inside the mound by this means, had the effect of reclaiming a considerable extent of fertile carse land, which, at the time of Southey's visit,—though the work had only been completed the year before,—was already under profitable cultivation. The principal use of the mound, however, was in giving support to the fine broad road which ran along its summit, and thus completed the communication with the country to the north. Southey speaks in terms of high admiration of "the simplicity, the beauty, and utility of this great work."

This was the furthest limit of their journey, and the travellers retraced their steps southward, halting at Clashmore Inn:—

"At breakfast," says Southey, "was a handsome set of Worcester china. Upon noticing it to Mr. Telford, he told me that before these roads were made, he fell in with some people from Worcestershire near the Ord of Caithness, on their way northward with a cart load of crockery, which they got over the mountains as best they could; and, when they had sold all their ware, they laid out the money in black cattle, which they then drove to the south."

The rest of Southey's journal is mainly occupied with a description of the scenery of the Caledonian Canal, and the principal difficulties encountered in the execution of the works, which were still in active progress. He was greatly struck with the flight of locks at the south end of the Canal, where it enters Loch Eil near Corpach:—

"There being no pier yet formed," he says, "we were carried to and from the boats on men's shoulders. We landed close to the sea shore. A sloop was lying in the fine basin above, and the canal was full as far as the Staircase, a name given to the eight successive locks. Six of these were full and overflowing; and then we drew near enough to see persons walking over the lock-gates. It had more the effect of a scene in a pantomime than of anything in real life. The rise from lock to lock is eight feet,—sixty-four, therefore, in all. The length of the locks, including the gates and abutments at both ends, is 500 yards;—the greatest piece of such masonry in the world, and the greatest work of the kind beyond all comparison.

"A panorama painted from this place would include the highest mountain in Great Britain, and its greatest work of art. That work is one of which the magnitude and importance become apparent, when considered in relation to natural objects. The Pyramids would appear insignificant in such a situation, for in them we should perceive only a vain attempt to vie with greater things. But here we see the powers of nature brought to act upon a great scale, in subservience to the purposes of men; one river created, another (and that a huge mountain-stream) shouldered out of its place, and art and order assuming a character of sublimity. Sometimes a beck is conducted under the canal, and passages called culverts serve as a roadway for men and beasts. We walked through one of these, just lofty enough for a man of my stature to pass through with his hat on. It had a very singular effect to see persons emerging from this dark, long, narrow vault. Sometimes a brook is taken in; a cesspool is then made to receive what gravel it may bring down after it has passed this pool, the water flowing through three or four little arches, and then over a paved bed and wall of masonry into the canal. These are called in-takes, and opposite them an outlet is sometimes made for the waters of the canal, if they should be above their proper level; or when the cross-stream may bring down a rush. These outlets consist of two inclined planes of masonry, one rising from the canal with a pavement or waste weir between them; and when the cross-stream comes down like a torrent, instead of mingling with the canal, it passes straight across. But these channels would be insufficient for carrying off the whole surplus waters in time of floods. At one place, therefore, there are three sluices by which the whole canal from the Staircase to the Regulating Lock (about six miles) can be lowered a foot in an hour. The sluices were opened that we might see their effect. We went down the bank, and made our way round some wet ground till we got in front of the strong arch into which they open. The arch is about 25 feet high, of great strength, and built upon the rock. What would the Bourbons have given for such a cascade at Versailles? The rush and the spray, and the force of the water, reminded me more of the Reichenbach than of any other fall. That three small sluices, each only 4 feet by 3 feet, should produce an effect which brought the mightiest of the Swiss waterfalls to my recollection, may appear incredible, or at least like an enormous exaggeration. But the prodigious velocity with which the water is forced out, by the pressure above, explains the apparent wonder. And yet I beheld it only in half its strength; the depth above being at this time ten feet, which will be twenty when the canal is completed. In a few minutes a river was formed of no inconsiderable breadth, which ran like a torrent into the Lochy.

"On this part of the canal everything is completed, except that the iron bridges for it, which are now on their way, are supplied by temporary ones. When the middle part shall be finished, the Lochy, which at present flows in its own channel above the Regulating Lock, will be dammed there, and made to join the Speyne by a new cut from the lake. The cut is made, and a fine bridge built over it. We went into the cut and under the bridge, which is very near the intended point of junction. The string-courses were encrusted with stalactites in a manner singularly beautiful. Under the arches a strong mound of solid masonry is built to keep the water in dry seasons at a certain height; but in that mound a gap is left for the salmon, and a way made through the rocks from the Speyne to this gap, which they will soon find out."

Arrived at Dumbarton, Southey took leave of John Mitchell, who had accompanied him throughout the tour, and for whom he seems to have entertained the highest admiration:—

"He is indeed," says Southey, "a remarkable man, and well deserving to be remembered. Mr. Telford found him a working mason, who could scarcely read or write. But his good sense, his excellent conduct, his steadiness and perseverance have been such, that he has been gradually raised to be Inspector of all these Highland roads which we have visited, and all of which are under the Commissioners' care—an office requiring a rare union of qualities, among others inflexible integrity, a fearless temper, and an indefatigable frame. Perhaps no man ever possessed these requisites in greater perfection than John Mitchell. Were but his figure less Tartarish and more gaunt, he would be the very 'Talus' of Spenser. Neither frown nor favour, in the course of fifteen years, have ever made him swerve from the fair performance of his duty, though the lairds with whom he has to deal have omitted no means of making him enter into their views, and to do things or leave them undone, as might suit their humour or interest. They have attempted to cajole and to intimidate him alike in vain. They have repeatedly preferred complaints against him in the hope of getting him removed from his office, and a more flexible person appointed in his stead; and they have not unfrequently threatened him with personal violence. Even his life has been menaced. But Mitchell holds right on. In the midst of his most laborious life, he has laboured to improve himself with such success, that he has become a good accountant, makes his estimates with facility, and carries on his official correspondence in an able and highly intelligent manner. In the execution of his office he travelled last year not less than 8800 miles, and every year he travels nearly as much. Nor has this life, and the exposure to all winds and weathers, and the temptations either of company or of solicitude at the houses at which he puts up, led him into any irregularities. Neither has his elevation in the slightest degree inflated him. He is still the same temperate, industrious, modest, unassuming man, as when his good qualities first attracted Mr. Telford's notice."

Southey concludes his journal at Longtown, a little town just across the Scotch Border, in the following words:—

"Here we left Mr. Telford, who takes the mail for Edinburgh. This parting company, after the thorough intimacy which a long journey produces between fellow-travellers who like each other, is a melancholy thing. A man more heartily to be liked, more worthy to be esteemed and admired, I have never fallen in with; and therefore it is painful to think how little likely it is that I shall ever see much of him again,—how certain that I shall never see *so* much. Yet I trust that he will not forget his promise of one day making Keswick in his way to and from Scotland."

Before leaving the subject of Telford's public works in the Highlands, it may be mentioned that 875 miles of new roads were planned by him, and executed under his superintendence, at an expense of 454,189*l.*, of which about one-half was granted by Parliament, and the remainder was raised by the localities benefited. Besides the new roads, 255 miles of the old military roads were taken in charge by him, and in many cases reconstructed and greatly improved. The bridges erected in connexion with these roads were no fewer than twelve hundred. Telford also between the year 1823 and the close of his life, built forty-two Highland churches in districts formerly unprovided with them, and capable of accommodating some 22,000 persons.

Down to the year 1854, the Parliamentary grant of 5000*l.* a year charged upon the Consolidated Fund to meet assessments and tolls of the Highland roads, amounting to about 7500*l.* a year, was transferred to the annual Estimates, when it became the subject of annual revision; and a few years since the grant was suddenly extinguished by an adverse vote of the House of Commons. The Board of Commissioners had, therefore, nothing left but to deliver over the roads to the several local authorities, and the harbours to the proprietors of the adjacent lands, and to present to Parliament a final account of their work and its results. Reviewing the whole, they say that the operations of the Commission have been most beneficial to the country concerned. They "found it barren and uncultivated, inhabited by heritors without capital or enterprise, and by a poor and ill-employed peasantry, and destitute of trade, shipping, and manufactures. They leave it with wealthy proprietors, a profitable agriculture, a thriving population, and active industry; furnishing now its fair proportion of taxes to the national exchequer, and helping by its improved agriculture to meet the ever-increasing wants of the populous south."

CHAPTER XV.

MR. TELFORD'S LATER YEARS—HIS DEATH AND CHARACTER.

WHEN Mr. Telford had occasion to visit London on business during the early period of his career, his quarters were at the Salopian Coffee House, now the Ship Hotel, at Charing Cross. It is probable that his Shropshire connections led him in the first instance to the 'Salopian;' but the situation being near to the Houses of Parliament, and in many respects convenient for the purposes of his business, he continued to live there for no less a period than twenty-one years. During that time the Salopian became a favourite resort of engineers; and not only Telford's provincial associates, but numerous visitors from abroad (where his works attracted even more attention than they did in England) took up their quarters there. Several apartments were specially reserved for Telford's exclusive use, and he could always readily command any additional accommodation for purposes of business or hospitality.

The successive landlords of the Salopian came to regard the engineer as a fixture, and even bought and sold him from time to time with the goodwill of the business. When he at length resolved, on the persuasion of his friends, to take a house of his own, and gave notice of his intention of leaving, the landlord, who had but recently entered into possession, almost stood aghast. "What! leave the house!" said he; "Why, Sir, I have just paid 750*l.* for you!" On explanation it appeared that this price had actually been paid by him to the outgoing landlord, on the assumption that Mr. Telford was a fixture of the hotel; the previous tenant having paid 450*l.* for him; the increase in the price marking very significantly the growing importance of the engineer's position. There was, however, no help for the disconsolate landlord, and Telford left the Salopian to take possession of his new house at 24, Abingdon Street. Labelye, the engineer of Westminster Bridge, had formerly occupied the dwelling; and, at a subsequent period, Sir William Chambers, the architect of Somerset House. Telford used to take much pleasure in pointing out to his visitors the painting of Westminster Bridge, impanelled in the wall over the parlour mantelpiece, made for Labelye by an Italian artist whilst the bridge works were in progress. In that house Telford continued to live until the close of his life.

One of the subjects in which he took much interest during his later years was the establishment of the Institute of Civil Engineers. In 1818 a Society had been formed, consisting principally of young men educated to civil and mechanical engineering, who occasionally met to discuss matters of interest relating to their profession. As early as the time of Smeaton, a social meeting of engineers was occasionally held at an inn in Holborn, which was discontinued in 1792, in consequence of some personal differences amongst the members. It was revived in the following year, under the auspices of Mr. Jessop, Mr. Naylor, Mr. Rennie, and Mr. Whitworth, and joined by other gentlemen of scientific distinction. They were accustomed to dine together every fortnight at the Crown and Anchor in the Strand, spending the evening in conversation on engineering subjects. But as the numbers and importance of the profession increased, the desire began to be felt, especially among the junior members of the profession, for an institution of a more enlarged character. Hence the movement above alluded to, which led to an invitation being given to Mr. Telford to accept the office of President of the proposed Engineers' Institute. To this he consented, and entered upon the duties of the office on the 21st of March, 1820.*

* In his inaugural address to the members on taking the chair, the President pointed out that the principles of the Institution rested

During the remainder of his life, Mr. Telford continued to watch over the progress of the Society, which gradually grew in importance and usefulness. He supplied it with the nucleus of a reference library, now become of great value to its members. He established the practice of recording the proceedings,* minutes of discussions, and substance of the papers read, which has led to the accumulation, in the printed records of the Institute, of a vast body of information as to engineering practice. In 1828 he exerted himself strenuously and successfully in obtaining a Charter of Incorporation for the Society; and finally, at his death, he left the Institute their first bequest of 2000*l*., together with many valuable books, and a large collection of documents which had been subservient to his own professional labours.

In the distinguished position which he occupied, it was natural that Mr. Telford should be called upon, as he often was, towards the close of his life, to give his opinion and advice as to projects of public importance. Where strongly

on the practical efforts and unceasing perseverance of the members themselves. "In foreign countries," he said, "similar establishments are instituted by government, and their members and proceedings are under their control; but here, a different course being adopted, it becomes incumbent on each individual member to feel that the very existence and prosperity of the Institution depend, in no small degree, on his personal conduct and exertions; and my merely mentioning the circumstance will, I am convinced, be sufficient to command the best efforts of the present and future members."

* We are informed by Joseph Mitchell, Esq., C.E., of the origin of this practice. Mr. Mitchell was a pupil of Mr. Telford's, living with him in his house at 24, Abingdon Street. It was the engineer's custom to have a dinner party every Tuesday, after which his engineering friends were invited to accompany him to the Institution, the meetings of which were then held on Tuesday evenings in a house in Buckingham Street, Strand. The meetings did not usually consist of more than from twenty to thirty persons. Mr. Mitchell took notes of the conversations which followed the reading of the papers. Mr. Telford afterwards found his pupil extending the notes, on which he asked permission to read them, and was so much pleased that he took them to the next meeting and read them to the members. Mr. Mitchell was then formally appointed reporter of conversations to the Institute; and the custom having been continued, a large mass of valuable practical information has thus been placed on record.

conflicting opinions were entertained on any subject, his help was occasionally found most valuable; for he possessed great tact and suavity of manner, which often enabled him to reconcile opposing interests when they stood in the way of important enterprises.

In 1828 he was appointed one of the commissioners to investigate the subject of the supply of water to the metropolis, in conjunction with Dr. Roget and Professor Brande, and the result was the very able report published in that year. Only a few months before his death, in 1834, he prepared and sent in an elaborate separate report, containing many excellent practical suggestions, which had the effect of stimulating the efforts of the water companies, and eventually leading to great improvements.

On the subject of roads, Telford continued to be the very highest authority, his friend Southey jocularly styling him the "Colossus of Roads." The Russian Government frequently consulted him with reference to the new roads with which that great empire was being opened up. The Polish road from Warsaw to Briesc, on the Russian frontier, 120 miles in length, was constructed after his plans, and it remains, we believe, the finest road in the Russian dominions to this day.

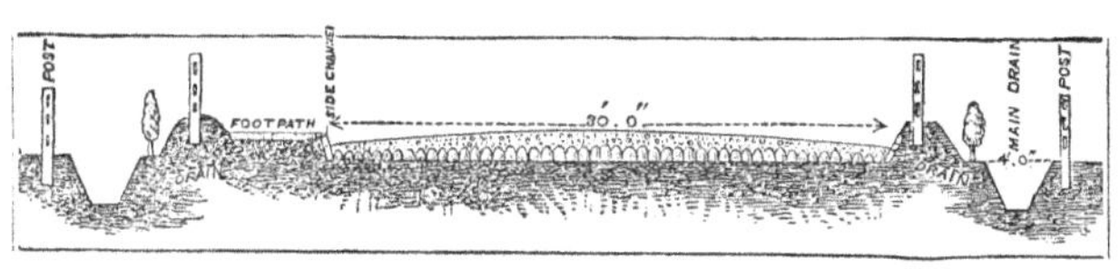

SECTION OF POLISH ROAD.

He was consulted by the Austrian Government on the subject of bridges as well as roads. Count Széchenyí recounts the very agreeable and instructive interview which he had with Telford when he called to consult him as to the bridge proposed to be erected across the Danube, between the towns of Buda and Pesth. On a suspension bridge being suggested by the English engineer, the

Count, with surprise, asked if such an erection was *possible* under the circumstances he had described? "We do not consider anything to be impossible," replied Telford; "impossibilities exist chiefly in the prejudices of mankind, to which some are slaves, and from which few are able to emancipate themselves and enter on the path of truth." But supposing a suspension bridge were not deemed advisable under the circumstances, and it were considered necessary altogether to avoid motion, "then," said he, "I should recommend you to erect a cast iron bridge of three spans, each 400 feet; such a bridge will have no motion, and though half the world lay a wreck, it would still stand." * A suspension bridge was eventually resolved upon. It was constructed by one of Mr. Telford's ablest pupils, Mr. Tierney Clark, between the years 1839 and 1850, and is justly regarded as one of the greatest triumphs of English engineering, the Buda-Pesth people proudly declaring it to be "the eighth wonder of the world."

At a time when speculation was very rife—in the year 1825—Mr. Telford was consulted respecting a grand scheme for cutting a canal across the Isthmus of Darien; and about the same time he was employed to resurvey the line for a ship canal—which had before occupied the attention of Whitworth and Rennie—between Bristol and the English Channel. But although he gave great attention to this latter project, and prepared numerous plans and reports upon it, and although an Act was actually passed enabling it to be carried out, the scheme was eventually abandoned, like the preceding ones with the same object, for want of the requisite funds.

Our engineer had a perfect detestation of speculative jobbing in all its forms, though on one occasion he could not help being used as an instrument by schemers. A public company was got up at Liverpool, in 1827, to form a broad and deep ship canal, of about seven miles in length,

* Supplement to Weale's 'Bridges,' Count Széchenyi's Report, p. 18.

from opposite Liverpool to near Helbre Isle, in the estuary of the Dee; its object being to enable the shipping of the port to avoid the variable shoals and sand-banks which obstruct the entrance to the Mersey. Mr. Telford entered on the project with great zeal, and his name was widely quoted in its support. It appeared, however, that one of its principal promoters, who had secured the right of pre-emption of the land on which the only possible entrance to the canal could be formed on the northern side, suddenly closed with the corporation of Liverpool, who were opposed to the plan, and "sold" his partners as well as the engineer for a large sum of money. Telford, disgusted at being made the instrument of an apparent fraud upon the public, destroyed all the documents relating to the scheme, and never afterwards spoke of it except in terms of extreme indignation.

About the same time, the formation of locomotive railways was extensively discussed, and schemes were set on foot to construct them between several of the larger towns. But Mr. Telford was now about seventy years old; and, desirous of limiting the range of his business rather than extending it, he declined to enter upon this new branch of engineering. Yet, in his younger days, he had surveyed numerous lines of railway—amongst others, one as early as the year 1805, from Glasgow to Berwick, down the vale of the Tweed. A line from Newcastle-on-Tyne to Carlisle was also surveyed and reported on by him some years later; and the Stratford and Moreton Railway was actually constructed under his direction. He made use of railways in all his large works of masonry, for the purpose of facilitating the haulage of materials to the points at which they were required to be deposited or used. There is a paper of his on the Inland Navigation of the County of Salop, contained in 'The Agricultural Survey of Shropshire,' in which he speaks of the judicious use of railways, and recommends that in all future surveys "it be an instruction to the engineers that they do examine the county with a view of introducing iron railways wherever difficulties may occur with regard to the making of navigable canals." When the project of the Liverpool and Manchester Railway was started, we are informed that he was offered the appointment of engineer; but he declined, partly because of his advanced age, but also out of a feeling of duty to his employers, the Canal Companies, stating that he could not lend his name to a scheme which, if carried out, must so materially affect their interests.

Towards the close of his life, he was afflicted by deafness, which made him feel exceedingly uncomfortable in mixed society. Thanks to a healthy constitution, unimpaired by excess and invigorated by active occupation, his working powers had lasted longer than those of most men. He was still cheerful, clear-headed, and skilful in the arts of his profession, and felt the same pleasure in useful work that he had ever done. It was, therefore, with difficulty that he could reconcile himself to the idea of retiring from the field of honourable labour, which he had so long occupied, into a state of comparative inactivity. But he was not a man who could be idle, and he determined, like his great predecessor Smeaton, to occupy the remaining years of his life in arranging his engineering papers for publication. Vigorous though he had been, he felt that the time was shortly approaching when the wheels of life must stand still altogether. Writing to a friend at Langholm, he said, "Having now being occupied for about seventy-five years in incessant exertion, I have for some time past arranged to decline the contest; but the numerous works in which I am engaged have hitherto prevented my succeeding. In the mean time I occasionally amuse myself with setting down in what manner a long life has been laboriously, and I hope usefully, employed." And again, a little later, he writes: "During the last twelve months I have had several rubs; at seventy-seven they tell more seriously than formerly, and call for less exertion and require greater precautions. I fancy that few of my age belonging to the valley of the Esk remain in the land of the living." *

One of the last works on which Mr. Telford was professionally consulted was at the instance of the Duke of Wellington—not many years younger than himself, but of equally vigorous intellectual powers—as to the improvement of Dover Harbour, then falling rapidly to decay. The long-continued south-westerly gales of 1833-4 had the effect of rolling an immense quantity of shingle up Channel towards that port, at the entrance to which it became deposited in unusual quantities, so as to render it at times altogether inaccessible. The Duke, as a military man, took a more than ordinary interest in the improvement of Dover, as the military and naval station nearest to the French coast; and it fell to him as Lord Warden of the Cinque Ports to watch over the preservation of the harbour, situated at a point in the English Channel which he regarded as of great strategic importance in the event of a continental war. He therefore desired Mr. Telford to visit the place and give his opinion as to the most advisable mode of procedure with a view to improving the harbour. The result was a report, in which the engineer recommended a plan of sluicing, similar to that adopted by Mr. Smeaton at Ramsgate, which was afterwards carried out with considerable success by Mr. James Walker, C.E.

This was his last piece of professional work. A few months later he was laid up by bilious derangement of a serious character, which recurred with increased violence towards the close of the year; and on the 2nd of September, 1834, Thomas Telford closed his useful and honoured career, at the advanced age of seventy-seven. With that absence of ostentation which characterised him through life, he directed that his remains should be laid, without ceremony, in the burial ground of the parish

* Letter to Mrs. Little, Langholm, 28th August, 1833.

church of St. Margaret's, Westminster. But the members of the Institute of Civil Engineers, who justly deemed him their benefactor and chief ornament, urged upon his executors the propriety of interring him in Westminster Abbey.

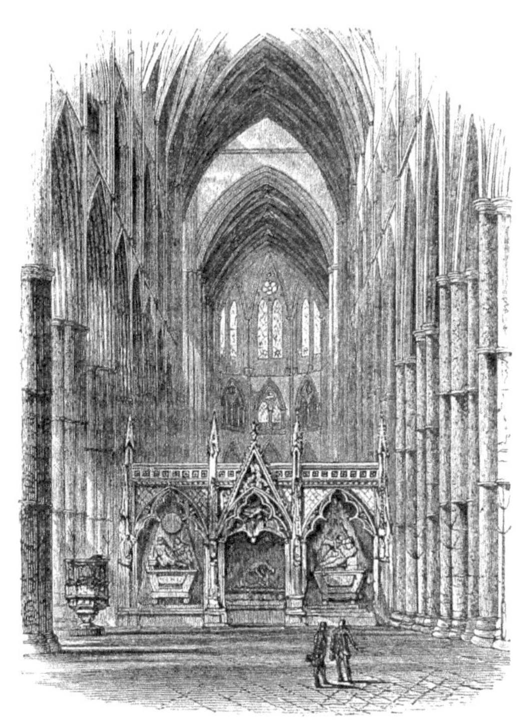

TELFORD'S BURIAL PLACE IN WESTMINSTER ABBEY.
[By Percival Skelton.]

He was buried there accordingly, near the middle of the nave; where the letters, "Thomas Telford, 1834," mark the place beneath which he lies.* The adjoining stone

* A statue of him, by Bailey, has since been placed in the east aisle of the north transept, known as the Islip Chapel. It is considered a fine work, but its effect is quite lost in consequence of the

bears the inscription, "Robert Stephenson, 1859," that engineer having during his life expressed the wish that his body should be laid near that of Telford; and the son of the Killingworth engineman thus sleeps by the side of the son of the Eskdale shepherd.

It was a long, a successful, and a useful life which thus ended. Every step in his upward career, from the poor peasant's hut in Eskdale to Westminster Abbey, was nobly and valorously won. The man was diligent and conscientious; whether as a working mason hewing stone blocks at Somerset House, as a foreman of builders at Portsmouth, as a road surveyor at Shrewsbury, or as an engineer of bridges, canals, docks, and harbours. The success which followed his efforts was thoroughly well-deserved. He was laborious, pains-taking, and skilful; but, what was better, he was honest and upright. He was a most reliable man; and hence he came to be extensively trusted. Whatever he undertook, he endeavoured to excel in. He would be a first-rate hewer, and he became one. He was himself accustomed to attribute much of his success to the thorough way in which he had mastered the humble beginnings of this trade. He was even of opinion that the course of manual training he had undergone, and the drudgery, as some would call it, of daily labour—first as an apprentice, and afterwards as a journeyman mason—had been of greater service to him than if he had passed through the curriculum of a University.

Writing to his friend, Miss Malcolm, respecting a young man who desired to enter the engineering profession, he in the first place endeavoured to dissuade the lady from encouraging the ambition of her *protégé*, the profession being overstocked, and offering very few prizes in proportion to

crowded state of the aisle, which has very much the look of a sculptor's workshop. The subscription raised for the purpose of erecting the statue was 1000*l.*, of which 200*l.* was paid to the Dean for permission to place it within the Abbey.

the large number of blanks. "But," he added, "if civil engineering, notwithstanding these discouragements, is still preferred, I may point out that the way in which both Mr. Rennie and myself proceeded, was to serve a regular apprenticeship to some practical employment—he to a millwright, and I to a general house-builder. In this way we secured the means, by hard labour, of earning a subsistence; and, in time, we obtained by good conduct the confidence of our employers and the public; eventually rising into the rank of what is called Civil Engineering. This is the true way of acquiring practical skill, a thorough knowledge of the materials employed in construction, and last, but not least, a perfect knowledge of the habits and dispositions of the workmen who carry out our designs. This course, although forbidding to many a young person, who believes it possible to find a short and rapid path to distinction, is proved to be otherwise by the two examples I have cited. For my own part, I may truly aver that 'steep is the ascent, and slippery is the way.'" *

That Mr. Telford was enabled to continue to so advanced an age employed on laborious and anxious work, was no doubt attributable in a great measure to the cheerfulness of his nature. He was, indeed, a most happy-minded man. It will be remembered that, when a boy, he had been known in his valley as "Laughing Tam." The same disposition continued to characterise him in his old age. He was playful and jocular, and rejoiced in the society of children and young people, especially when well-informed and modest. But when they pretended to acquirements they did not possess, he was quick to detect and see through them. One day a youth expatiated to him in very large terms about a friend of his, who had done this and that, and made so and so, and could do all manner of wonderful

* Letter to Miss Malcolm, Burnfoot, Langholm, dated 7th October, 1830.

things. Telford listened with great attention, and when the youth had done, he quietly asked, with a twinkle in his eye, "Pray, can your friend lay eggs?"

When in society he gave himself up to it, and thoroughly enjoyed it. He did not sit apart, a moody and abstracted "lion;" nor desire to be regarded as "the great engineer," pondering new Menai Bridges; but he appeared in his natural character of a simple, intelligent, cheerful companion; as ready to laugh at his own jokes as at other people's; and he was as communicative to a child as to any philosopher of the party.

Robert Southey, than whom there was no better judge of a loveable man, said of him, "I would go a long way for the sake of seeing Telford and spending a few days in his company." Southey, as we have seen, had the best opportunities of knowing him well; for a long journey together extending over many weeks, is, probably, better than anything else, calculated to bring out the weak as well as the strong points of a friend: indeed, many friendships have completely broken down under the severe test of a single week's tour. But Southey on that occasion firmly cemented a friendship which lasted until Telford's death. On one occasion the latter called at the poet's house, in company with Sir Henry Parnell, when engaged upon the survey of one of his northern roads. Unhappily Southey was absent at the time; and, writing about the circumstance to a correspondent, he said, "This was a mortification to me, inasmuch as I owe Telford every kind of friendly attention, and like him heartily."

Campbell, the poet, was another early friend of our engineer; and the attachment seems to have been mutual. Writing to Dr. Currie, of Liverpool, in 1802, Campbell says: "I have become acquainted with Telford the engineer, 'a fellow of infinite humour,' and of strong enterprising mind. He has almost made me a bridge-builder already; at least he has inspired me with new sensations of interest in the improvement and ornament of our country. Have you seen his plan of London Bridge? or his scheme for a new canal in the North Highlands, which will unite, if put in effect, our Eastern and Atlantic commerce, and render Scotland the very emporium of navigation? Telford is a most useful cicerone in London. He is so universally acquainted, and so popular in his manners, that he can introduce one to all kinds of novelty, and all descriptions of interesting society." Shortly after, Campbell named his first son after Telford, who stood godfather for the boy. Indeed, for many years, Telford played the part of Mentor to the young and impulsive poet, advising him about his course in life, trying to keep him steady, and holding him aloof as much as possible from the seductive allurements of the capital. But it was a difficult task, and Telford's numerous engagements necessarily left the poet at many seasons very much to himself. It appears that they were living together at the Salopian when Campbell composed the first draft of his poem of Hohenlinden; and several important emendations made in it by Telford were adopted by Campbell. Although the two friends pursued different roads in life, and for many years saw little of each other, they often met again, especially after Telford took up his abode at his house in Abingdon Street, where Campbell was a frequent and always a welcome guest.

When engaged upon his surveys, our engineer was the same simple, cheerful, laborious man. While at work, he gave his whole mind to the subject in hand, thinking of nothing else for the time; dismissing it at the close of each day's work, but ready to take it up afresh with the next day's duties. This was a great advantage to him as respected the prolongation of his working faculty. He did not take his anxieties to bed with him, as many do, and rise up with them in the morning; but he laid down the load at the end of each day, and resumed it all the more cheerfully when refreshed and invigorated by natural rest. It was only while the engrossing anxieties connected with the suspension of the chains of Menai Bridge were weighing heavily upon his mind, that he could not sleep; and then, age having stolen upon him, he felt the strain almost more than he could bear. But that great anxiety once fairly over, his spirits speedily resumed their wonted elasticity.

When engaged upon the construction of the Carlisle and Glasgow road, he was very fond of getting a few of the "navvy men," as he called them, to join him at an ordinary at the Hamilton Arms Hotel, Lanarkshire, each paying his own expenses. On such occasions Telford would say that, though he could not drink, yet he would carve and draw corks for them. One of the rules he laid down was that no business was to be introduced from the moment they sat down to dinner. All at once, from being the plodding, hard-working engineer, with responsibility and thought in every feature, Telford unbended and relaxed, and became the merriest and drollest of the party. He possessed a great fund of anecdote available for such occasions, had an extraordinary memory for facts relating to persons and families, and the wonder to many of his auditors was, how in all the world a man living in London should know so much better about their locality and many of its oddities than they did themselves.

In his leisure hours at home, which were but few, he occupied himself a good deal in the perusal of miscellaneous literature, never losing his taste for poetry. He continued to indulge in the occasional composition of verses until a comparatively late period of his life; one of his most successful efforts being a translation of the 'Ode to May,' from Buchanan's Latin poems, executed in a very tender and graceful manner. That he might be enabled to peruse engineering works in French and German, he prosecuted the study of those languages, and with such success that he was shortly able to read them with comparative ease. He occasionally occupied himself in literary composition on subjects connected with his profession. Thus he wrote for the Edinburgh Encyclopedia, conducted by his friend Sir David (then Dr.) Brewster, the elaborate and able articles on Architecture, Bridge-building, and Canal-making. Besides his contributions to that work, he advanced a considerable sum of money to aid in its publication, which remained a debt due to his estate at the period of his death.

Notwithstanding the pains that Telford took in the course of his life to acquire a knowledge of the elements of natural science, it is somewhat remarkable to find him holding acquirements in mathematics so cheap. But probably this is to be accounted for by the circumstance of his education being entirely practical, and mainly self-acquired. When a young man was on one occasion recommended to him as a pupil because of his proficiency in mathematics, the engineer expressed the opinion that such acquirements were no recommendation. Like Smeaton, he held that deductions drawn from theory were never to be trusted; and he placed his reliance mainly on observation, experience, and carefully-conducted experiments. He was also, like most men of strong practical sagacity, quick in mother wit, and arrived rapidly at conclusions, guided by a sort of intellectual instinct which can neither be defined nor described.*

Although occupied as a leading engineer for nearly forty years—having certified contractors' bills during that time amounting to several millions sterling—he died in com-

* Sir David Brewster observes on this point: "It is difficult to analyse that peculiar faculty of mind which directs a successful engineer who is not guided by the deductions of the exact sciences; but it must consist mainly in the power of observing the effects of natural causes acting in a variety of circumstances; and in the judicious application of this knowledge to cases when the same causes come into operation. But while this sagacity is a prominent feature in the designs of Mr. Telford, it appears no less distinctly in the choice of the men by whom they were to be practically executed. His quick perception of character, his honesty of purpose, and his contempt for all other acquirements,—save that practical knowledge and experience which was best fitted to accomplish, in the best manner, the object he had in view,—have enabled him to leave behind him works of inestimable value, and monuments of professional celebrity which have not been surpassed either in Britain or in Europe."—'Edinburgh Review,' vol. lxx. p. 46.

paratively moderate circumstances. Eminent constructive ability was not very highly remunerated in Telford's time, and he was satisfied with a rate of pay which even the smallest "M. I. C. E." would now refuse to accept. Telford's charges were, however, perhaps too low; and a deputation of members of the profession on one occasion formally expostulated with him on the subject.

Although he could not be said to have an indifference for money, he yet estimated it as a thing worth infinitely less than character; and every penny that he earned was honestly come by. He had no wife,* nor family, nor near relations to provide for,—only himself in his old age. Not being thought rich, he was saved the annoyance of being haunted by toadies or pestered by parasites. His wants were few, and his household expenses small; and though he entertained many visitors and friends, it was in a quiet way and on a moderate scale. The small regard he had for personal dignity may be inferred from the fact, that to the last he continued the practice, which he had learnt when a working mason, of darning his own stockings.†

* It seems singular that with Telford's great natural powers of pleasing, his warm social temperament, and his capability of forming ardent attachments for friends, many of them women, he should never have formed an attachment of the heart. Even in his youthful and poetical days, the subject of love, so frequently the theme of boyish song, is never alluded to; while his school friendships are often recalled to mind, and, indeed, made the special subject of his verse. It seems odd to find him, when at Shrewsbury—a handsome fellow, with a good position, and many beautiful women about him—addressing his friend, the blind schoolmaster at Langholm, as his "Stella"!

† Mr. Mitchell says: "He lived at the rate of about 1200*l.* a year. He kept a carriage, but no horses, and used his carriage principally for making his journeys through the country on business. I once accompanied him to Bath and Cornwall, when he made me keep an accurate journal of all I saw. He used to lecture us on being independent, even in little matters, and not ask servants to do for us what we might easily do for ourselves. He carried in his pocket a small book containing needles, thread, and buttons, and on an emergency was always ready to put in a stitch. A curious habit he had of mending his stockings, which I suppose he acquired when a working mason. He would not permit his housekeeper to touch them, but after his work at night, about nine or half-

Telford nevertheless had the highest idea of the dignity of his profession; not because of the money it would produce, but of the great things it was calculated to accomplish. In his most confidential letters we find him often expatiating on the noble works he was engaged in designing or constructing, and the national good they were calculated to produce, but never on the pecuniary advantages he himself was to derive from them. He doubtless prized, and prized highly, the reputation they would bring him; and, above all, there seemed to be uppermost in his mind, especially in the earlier part of his career, while many of his schoolfellows were still alive, the thought of "What will they say of this in Eskdale?" but as for the money results to himself, Telford seemed, to the close of his life, to regard them as of comparatively small moment.

During the twenty-one years that he acted as principal engineer for the Caledonian Canal, we find from the Parliamentary returns that the amount paid to him for his reports, detailed plans, and superintendence, was exactly 237*l.* a year. Where he conceived any works to be of great public importance, and he found them to be promoted by public-spirited persons at their own expense, he refused to receive any payment for his labour, or even repayment of the expenses incurred by him. Thus, while employed by the Government in the improvement of the Highland roads, he persuaded himself that he ought at the same time to promote the similar patriotic objects of the British Fisheries Society, which were carried out by voluntary subscription; and for many years he acted as their engineer, refusing to accept any remuneration whatever for his trouble.*

past, he would go up stairs, and take down a lot, and sit mending them with great apparent delight in his own room till bed-time. I have frequently gone in to him with some message, and found him occupied with this work."

* "The British Fisheries Society," adds Mr. Rickman, "did not suffer themselves to be entirely outdone in liberality, and shortly before his death they pressed upon Mr. Telford a very handsome gift of plate, which, being inscribed

Telford held the sordid money-grubber in perfect detestation. He was of opinion that the adulation paid to mere money was one of the greatest dangers with which modern society was threatened. "I admire commercial enterprise," he would say; "it is the vigorous outgrowth of our industrial life: I admire everything that gives it free scope, as, wherever it goes, activity, energy, intelligence—all that we call civilization—accompany it; but I hold that the aim and end of all ought not to be a mere bag of money, but something far higher and far better."

Writing once to his Langholm correspondent about an old schoolfellow, who had grown rich by scraping, Telford said: "Poor Bob L——! His industry and sagacity were more than counterbalanced by his childish vanity and silly avarice, which rendered his friendship dangerous, and his conversation tiresome. He was like a man in London, whose lips, while walking by himself along the streets, were constantly ejaculating 'Money! Money!' But peace to Bob's memory: I need scarcely add, confusion to his thousands!" Telford was himself most careful in resisting the temptations to which men in his position are frequently exposed; but he was preserved by his honest pride, not less than by the purity of his character. He invariably refused to receive anything in the shape of presents or testimonials from persons employed under him. He would not have even the shadow of an obligation stand in the way of his duty to those who employed him to watch over and protect their interests. During the many years that he was employed on public works, no one could ever charge him in the remotest degree with entering into a collusion with contractors. He looked upon such arrangements as degrading and infamous, and considered that they meant nothing less than an inducement to "scamping," which he would never tolerate.

with expressions of their thankfulness and gratitude towards him, he could not possibly refuse to accept."—'Life of Telford,' p. 283.

His inspection of work was most rigid. The security of his structures was not a question of money, but of character. As human life depended upon their stability, not a point was neglected that could ensure it. Hence, in his selection of resident engineers and inspectors of works, he exercised the greatest possible precautions; and here his observation of character proved of essential value. Mr. Hughes says he never allowed any but his most experienced and confidential assistants to have anything to do with exploring the foundations of buildings he was about to erect. His scrutiny into the qualifications of those employed about such structures extended to the subordinate overseers, and even to the workmen, insomuch that men whose general habits had before passed unnoticed, and whose characters had never been inquired into, did not escape his observation when set to work in operations connected with foundations.* If he detected a man who gave evidences of unsteadiness, inaccuracy, or carelessness, he would reprimand the overseer for employing such a person, and order him to be removed to some other part of the undertaking where his negligence could do no harm. And thus it was that Telford put his own character, through those whom he employed, into the various buildings which he was employed to construct.

But though Telford was comparatively indifferent about money, he was not without a proper regard for it, as a means of conferring benefits on others, and especially as a means of being independent. At the close of his life he had accumulated as much as, invested at interest, brought him in about 800*l*. a year, and enabled him to occupy the house in Abingdon Street in which he died. This was amply sufficient for his wants, and more than enough for his independence. It enabled him also to continue those secret acts of benevolence which constituted perhaps the

* Weale's 'Theory, Practice, and Architecture of Bridges,' vol. i.: 'Essay on Foundations of Bridges,' by T. Hughes, C.E., p. 33.

most genuine pleasure of his life. It is one of the most delightful traits in this excellent man's career to find him so constantly occupied in works of spontaneous charity, in quarters so remote and unknown that it is impossible the slightest feeling of ostentation could have sullied the purity of the acts. Among the large mass of Telford's private letters which have been submitted to us, we find frequent reference to sums of money transmitted for the support of poor people in his native valley. At new year's time he regularly sent remittances of from 30*l.* to 50*l.*, to be distributed by the kind Miss Malcolm of Burnfoot, and, after her death, by Mr. Little, the postmaster at Langholm; and the contributions thus so kindly made, did much to fend off the winter's cold, and surround with many small comforts those who most needed help, but were perhaps too modest to ask it.

Many of those in the valley of the Esk had known of Telford in his younger years as a poor barefooted boy; though now become a man of distinction, he had too much good sense to be ashamed of his humble origin; perhaps he even felt proud that, by dint of his own valorous and persevering efforts, he had been able to rise so much above it. Throughout his long life, his heart always warmed at the thought of Eskdale. He rejoiced at the honourable rise of Eskdale men as reflecting credit upon his "beloved valley." Thus, writing to his Langholm correspondent with reference to the honours conferred on the different members of the family of Malcolm, he said: "The distinctions so deservedly bestowed upon the Burnfoot family, establish a splendid era in Eskdale; and almost tempt your correspondent to sport his Swedish honours, which that grateful country has repeatedly, in spite of refusal, transmitted."*

It might be said that there was narrowness and provincialism in this; but when young men are thrown

* Letter to Mr. William Little, Langholm, 24th January, 1815.

into the world, with all its temptations and snares, it is well that the recollections of home and kindred should survive to hold them in the path of rectitude, and cheer them in their onward and upward course in life. And there is no doubt that Telford was borne up on many occasions by the thought of what the folks in the valley would say about him and his progress in life, when they met together at market, or at the Westerkirk porch on Sabbath mornings. In this light, provincialism or local patriotism is a prolific source of good, and may be regarded as among the most valuable and beautiful emanations of the parish life of our country. Although Telford was honoured with the titles and orders of merit conferred upon him by foreign monarchs, what he esteemed beyond them all was the respect and gratitude of his own countrymen; and, not least, the honour which his really noble and beneficent career was calculated to reflect upon "the folks of the nook," the remote inhabitants of his native Eskdale.

When the engineer proceeded to dispose of his savings by will, which he did a few months before his death, the distribution was a comparatively easy matter. The total amount of his bequeathments was 16,600*l.** About one-fourth of the whole he set apart for educational purposes, —2000*l.* to the Civil Engineers' Institute, and 1000*l.* each to the ministers of Langholm and Westerkirk, in trust for the parish libraries. The rest was bequeathed, in sums of from 200*l.* to 500*l.*, to different persons who had acted as clerks, assistants, and surveyors, in his various public

* Telford thought so little about money, that he did not even know the amount he died possessed of. It turned out that instead of 16,600*l.* it was about 30,000*l.*, so that his legatees had their bequests nearly doubled. For many years he had abstained from drawing the dividends on the shares which he held in the canals and other public companies in which he was concerned. At the money panic of 1825, it was found that he had a considerable sum lying in the hands of his London bankers at little or no interest, and it was only on the urgent recommendation of his friend, Sir P. Malcolm, that he invested it in government securities, then very low.

works; and to his intimate personal friends. Amongst these latter were Colonel Pasley, the nephew of his early benefactor; Mr. Rickman, Mr. Milne, and Mr. Hope, his three executors; and Robert Southey and Thomas Campbell, the poets. To both of these last the gift was most welcome. Southey said of his: "Mr. Telford has most kindly and unexpectedly left me 500*l.*, with a share of his residuary property, which I am told will make it amount in all to 850*l.* This is truly a godsend, and I am most grateful for it. It gives me the comfortable knowledge that, if it should please God soon to take me from this world, my family would have resources fully sufficient for their support till such time as their affairs could be put in order, and the proceeds of my books, remains, &c., be rendered available. I have never been anxious overmuch, nor ever taken more thought for the morrow than it is the duty of every one to take who has to earn his livelihood; but to be thus provided for at this time I feel to be an especial blessing." *

Among the most valuable results of Telford's bequests in his own district, was the establishment of the popular libraries at Langholm and Westerkirk, each of which now contains about 4000 volumes. That at Westerkirk had been originally instituted in the year 1792, by the miners employed to work an antimony mine (since abandoned) on the farm of Glendinning, within sight of the place where Telford was born. On the dissolution of the mining company, in 1800, the little collection of books was removed to Kirkton Hill; but on receipt of Telford's bequest, a special building was erected for their reception at Old Bentpath near the village of Westerkirk. The annual income derived from the Telford fund enabled additions of new volumes to be made to it from time to time; and its

* 'Selections from the Letters of Robert Southey,' vol. iv., p. 391. We may here mention that the last article which Southey wrote for the 'Quarterly' was his review of the 'Life of Telford.'

uses as a public institution were thus greatly increased. The books are exchanged once a month, on the day of the full moon; on which occasion readers of all ages and conditions,—farmers, shepherds, ploughmen, labourers, and their children,—resort to it from far and near, taking away with them as many volumes as they desire for the month's reading.

Thus there is scarcely a cottage in the valley in which good books are not to be found under perusal; and we are told that it is a common thing for the Eskdale shepherd to take a book in his plaid to the hill-side—a volume of Shakespeare, Prescott, or Macaulay—and read it there, under the blue sky, with his sheep and the green hills before him. And thus, so long as the bequest lasts, the good, great engineer will not cease to be remembered with gratitude in his beloved Eskdale.

THE END.

INDEX.

PRINTED BY W. CLOWES AND SONS, DUKE STREET, STAMFORD STREET, AND CHARING CROSS.